THE ANALOGUE IDYLL

Disconnection, Detox, and Departure from the Digital World

Edited by
A.R.E. Taylor

With a Foreword by
Trine Syvertsen

BRISTOL
UNIVERSITY
PRESS

First published in Great Britain in 2026 by

Bristol University Press
University of Bristol
1-9 Old Park Hill
Bristol
BS2 8BB
UK
t: +44 (0)117 374 6645
e: bup-info@bristol.ac.uk

Details of international sales and distribution partners are available at bristoluniversitypress.co.uk

DOI: 10.51952/9781529233728

British Library Cataloguing in Publication Data
A catalogue record for this book is available from the British Library

ISBN 978-1-5292-3370-4 hardcover
ISBN 978-1-5292-3371-1 ePub
ISBN 978-1-5292-3372-8 ePdf

Cover design: Liam Roberts Design
Front cover image: iStock/luza studios
Bristol University Press uses environmentally responsible print partners.
Printed and bound by CPI Group (UK) Ltd, Croydon, CR0 4YY

Bristol University Press' authorised representative in the European Union is:
Easy Access System Europe, Mustamäe tee 50, 10621 Tallinn, Estonia,
Email: gpsr.requests@easproject.com

FSC
www.fsc.org
MIX
Paper | Supporting responsible forestry
FSC® C013604

Contents

List of Figures and Tables

Figures

Tables

Notes on Contributors

Kristoffer Albris is Associate Professor at the Department of Anthropology and the Copenhagen Center for Social Data Science (SODAS), University of Copenhagen. His research focuses on attention in the digital age, computational ethnography, and data governance. He was vice-PI on the ERC-funded project DISTRACT: The Political Economy of Attention in Digitized Denmark.

Neil Ewen is Associate Professor of Media, Communications and Culture at the University of Exeter. His research interests lie broadly in the politics of contemporary media and culture. He is the co-editor of *Coming Back Stronger: 1990s Nostalgia in Contemporary Media Culture* (Bloomsbury Academic, 2026), *Capitalism, Crime and Media in the 21st Century* (Palgrave Macmillan, 2021), *First Comes Love: Power Couples, Celebrity Kinship, and Cultural Politics* (New York: Bloomsbury Academic, 2015) and special issues of *Television and New Media* on the TV show *Friends* (2018) and *Celebrity Studies* on populist political celebrities in contemporary Europe (2020). He edited the Cultural Report section of the journal *Celebrity Studies* from 2015–2025 and is a book series editor of *Games and Contemporary Culture* (Routledge). He is currently writing a book about football in film and digital media.

Malene Hornstrup Jespersen is a PhD fellow in Social Data Science at SODAS, University of Copenhagen. She holds a BSc in Anthropology and a MSc in IT and Cognition. Her research focuses on attentionality and digital disconnection practices, values and discourses. The PhD forms part of the ERC-funded project DISTRACT.

Annika Isfeldt is a PhD fellow at Educational Anthropology, Aarhus University. Her current research focuses on AI, ethics, futures, and humanitarian work. Prior to this, she was affiliated with the DISTRACT project, researching the concept of attentionality, digital detoxing, and political struggles over attention.

Linda Kopitz connects her professional experience as a Creative Director with her interdisciplinary academic research to explore the role of technology in everyday meaning-making. She is currently working as a Lecturer in Cross-Media Culture in the Netherlands and Germany, where her main research interests are architectural media, urban sustainability and the entanglement between real and virtual environments. Bringing together her academic and editorial work, she is an assistant editor for the *European Journal of Cultural Studies*.

Alexandra Kviat is Lecturer in Marketing and Consumption at the University of Bristol Business School. She works across the fields of sociology, cultural and media studies, human geography, and marketing theory. Her interdisciplinary research projects have explored the relationship between technology, space and consumption in the context of the hospitality, retail and leisure industries. Alexandra's work has been supported by the Leverhulme Trust, the Economic and Social Research Council, the University of Warwick Institute of Advanced Study and Chancellor's International Scholarship, and the Fulbright Program.

Dave McLaughlin is Lecturer in Human Geography at the University of East Anglia. He has a PhD in Geography from the University of Cambridge. His research uses more-than-representational theories to explore the affective force and agency of representations (digital and analogue) in the co-production of subjectivities, communities and landscapes. His recent publications include a study of the affective force of hikers' writings in co-producing identities and communities along America's Appalachian Trail (AT); and an exploration of the role of AT hikers' digital representations as unruly writings that reflect and enact the Appalachian Trail as an unruly landscape.

Paul O'Connor is a qualitative sociologist from the UK. He received his PhD from the University of Queensland studying Muslim youth in Hong Kong. His research explores religion, ethnicity, Hong Kong studies, and lifestyle sports. He is author of the two books: *Islam in Hong Kong*, with HKU Press (2012) and *Skateboarding and Religion*, with Palgrave (2020). He has lived and taught in Prague, and Hong Kong and holds an adjunct position at Hong Kong's Lignan University. His current work is interested in an emerging digital divide where disengagement is privileged. He currently lectures in Sociology and Anthropology at the University of Exeter.

Michael Palm is Associate Professor and Director of Cultural Studies in the Department of Communication at UNC-Chapel Hill. He is also affiliated faculty in the Department of American Studies, and he serves as President of UNC's chapter of the American Association of University Professors. His research and teaching focus on the history of everyday technology and the political economy of popular culture. His book, *Technologies of Consumer Labor: A History of Self-Service*, was published by Routledge in 2017. He has published solo and co-authored articles in journals such as *Media, Culture & Society*, *The European Journal of Cultural Studies*, and *The Journal of Popular Music Studies*. He is currently writing a book about the labour, logistics, and ecology of the revived economy for vinyl records.

Trine Syvertsen is Professor of Media Studies at the University of Oslo, working in the fields of media policy, history and use. She has published articles and books about digital disconnection and resistance including *Media Resistance: Protest, Dislike, Abstention* (Springer, 2017) and *Digital Detox: The Politics of Disconnecting* (Emerald Publishing, 2020). Together with colleagues she has edited the collection *The Digital Backlash and the Paradoxes of Disconnection* (Nordicom, 2024). She chairs the project Intrusive Media, Ambivalent Users and Digital Detox (Digitox) which runs online seminars about issues related to digital disconnection.

A.R.E. Taylor is an anthropologist and Senior Lecturer in Communications at the University of Exeter. He works at the intersection of social anthropology, infrastructure studies, and media and communication studies. His research concentrates on the material infrastructure and labour that underpin digital services. He is especially interested in forms of unanticipated disconnection from the internet, arising from the failure and breakdown of internet infrastructure. He is an Editor for the *Journal of Extreme Anthropology* and a founder of the *Cambridge Infrastructure Resilience Group*, a network of researchers exploring critical infrastructure protection in relation to global catastrophic risks.

Acknowledgements

I would like to thank the Culture-Media-Text Research Centre at the University of Winchester for their generous support of the analogue idyll symposium, which took place in August 2020. I am hugely grateful to all the symposium attendees for finding the time and energy to participate despite the widespread Zoom fatigue during the height of the COVID-19 pandemic. Many of those who joined the symposium contributed chapters to this volume. I want to thank all of the contributors for their patience over the last few years and for sticking with the project as the volume came together against the backdrop of the global pandemic, when many of us were simultaneously grieving the deaths of loved ones while leap-frogging from job-to-job, navigating the precarity of the neoliberal academy. I would like to thank Neil Ewen, Inga Bryden, Jennifer Cearns, Theodora Sutton, Marcus Leaning, and Jude Davies for supporting the project at various stages. Thanks also to the anonymous reviewers who praised the collection and provided invaluable feedback. At Bristol University Press, I would like to thank Paul Stevens and Ellen Mitchell for their patience and support. I would also like to thank Claire Bell for her careful copyediting work.

Foreword

Trine Syvertsen

I live in Norway, which was once ranked as the happiest country in the world (The World Happiness Report, 2016). But then its rank started dropping. In 2024, it had fallen to seventh place.

The Norwegian Government used to brag about the happy state. Happiness figured prominently in nation-branding. But what is there to brag about now?

Incidentally, the Government has just proclaimed that it aims for another world record: to make Norway 'the most digitalized country in the world' by 2030.

According to the Government action plan, digitalization 'will make everyday life easier and safer for citizens, make businesses more competitive and improve the public sector'. Norway will be 'safer and fairer', and citizens will get the 'opportunity to enjoy a good quality of life' (Ministry of Digitalisation and Public Governance, 2024).

Apparently, we do not need to be concerned about our (declining) happiness. Digitalization will achieve the same goal.

But not everyone is happy in the digital paradise.

For the five young women in central Oslo who have just launched a paper magazine called 'På' (Meaning 'On' or more metaphorically 'Being present'), it all started as a protest against META. By now, it has turned into a small movement. The magazine promotes offline and cultural events and is curating informal dinners for people to meet face-to-face. More people are attending every time. They say that some people see them as a 'cult', but they 'just want to encourage a more physical connection' (Utaker, 2025).

While these young women recreate 'the analogue idyll', those who are less digital are not as celebratory.

For the 70- to 90-year-olds who had invited me to speak at their 'debate club' in Southern Norway, the state of affairs was still largely analogue. When I arrived, the members of the 129-year-old club, which began as part of the fight for female suffrage, had just finished their annual meeting. Procedures were conducted, and minutes were recorded with pen and paper. The members had quite a few (unhappy) things to say about the digital society, but expressed little nostalgia for the communication channels of their youth.

The idyllic version of analogue is not for all.

We have plenty of terms to describe our current condition. We talk about the post-digital, post-virtual, and surveillance society. We observe a tech-lash and digital backlash. We record symptoms of digital resignation and distancing. We write about individuals who engage in a digital detox or disconnect from social media. We log off, go offline, or go off the grid when we've had too much. We record techno-nostalgic sentiments, retro media use, and slow media practices. We discuss whether any of this qualifies as media or digital resistance.

All these are useful and interesting concepts and have books and articles to back them up.

What can 'the analogue idyll' add to this myriad of terms?

'The analogue idyll' can be used to explore why some people prefer physical, tangible media. Paper books, board games, postcards, vinyl, typewriters, maps, compasses, and old phones are all props in the idyllic analogue universe. It does not matter if these are partly digital or depend on digital infrastructures. Analogue fans do not rely on strict categories for fandom.

But 'the analogue idyll' is not just about artefacts. As the excellent case studies in this collection illuminate, it helps explore selectivity, morality, temporality, and locality.

'The analogue idyll', just like the rural idyll, signals a high level of *selectivity*. Visiting an old farm and cooking over an open fire, with phones tucked away in backpacks, would definitely qualify. But wearing stiff old boots, sleeping on flea-infested straw mattresses, or wearing an old leaky raincoat is not part of it. Indeed, phone-free outdoor experiences often serve as catwalks for the newest, most high-tech gear. While some retro objects indicate a celebration of the rural or analogue idyll, others just indicate that you are poor or out of touch.

'The analogue idyll' helps explore *morality*. Many of those who practice digital disconnection and use analogue media emphasize that they are non-judgemental. But the point is still to establish a hierarchy of meaning and importance. Yet, propagating moral hierarchies is difficult in our day and age. 'We are not against technology' is the title of the newspaper article about the women who started the paper magazine in Oslo. JUST, SO YOU KNOW, I AM NOT AGAINST THE INTERNET, says the preface of every book about addictive technology and social media's colonization of everyday life.

There used to be people who were against technology and new media. People protested the cinema, comics, rock music, television, video games, and reality formats. But nowadays, tech has won, and previously hated media have become lovable, nostalgic objects.

'The analogue idyll' illuminates contested *temporality*. It belongs to a family of slow experiences. Since the slow food movement emerged in

Italy in the 1980s, 'slow' has been applied to domains such as work, travel, science, adventure, media and more. While digital media are seen as vehicles of globalization, commercialization, and acceleration, 'slow' connotes co-creation, low environmental impact and mindful presence. Walking is part of 'the analogue idyll', but perhaps not powerwalking or logging your walks on *Strava*.

'The analogue idyll' helps explain matters of *locality*. It illuminates people's relations to places and situations. Food and coffee, outdoors and nature, culture and music venues, and anything local can be infused with idyllic meaning. 'The analogue idyll' can be used to explore what it is in a place that invites slow immersion, presence and belonging, and why some places seem appropriate for writing a postcard to a friend.

The media love 'the analogue idyll'. Indeed, the phenomenon may be bigger in the media than in everyday life.

'The analogue idyll' is already an old trope in literature and popular culture. Despite the irony and the hipster vibes surrounding it, it is worth noting that it is usually a hopeful and optimistic trope.

In Bradbury's *Fahrenheit 451* (1953), hope lies in the 'book people', an illicit group living on abandoned railway tracks away from the screen-covered walls of private homes. Hope lies in owning a paper notebook and avoiding the telescreens in Orwell's *Nineteen Eighty-Four* (1949). It can be found on the islands where writers and scientists are exiled to write in Huxley's *Brave New World* (1932).

The main character in *Her* (Jonze, 2013) is a tragic and lonely figure. But we know he is our hero because he works for the firm 'Beautiful-handwritten-letters-dot-com'. He composes love letters for people who have sadly lost the ability to write their own.

Even the corrupt, rich people in the third season of *The White Lotus* (White, 2025), who mainly engage in drug-taking, deceitful sex, lying, fighting, and destruction, connect better with their nearest and dearest when their spa therapists remove their phones.

Indeed, at least in fiction, 'the analogue idyll' is the saviour of language, education, history, memory, nature, and even love.

References

Bradbury, R. (2013 [1953]) *Fahrenheit 451*. Simon and Schuster.

Huxley, A. (2006 [1932]) *Brave New World*. Harper Perennial.

Jonze, S. (2013) *Her*. Warner Bros. Pictures.

Ministry of Digitalisation and Public Governance (2024) *The Digital Norway of the Future, National Digitalisation Strategy 2024–2030*. Regjeringen.no, [online] 4 November, Available from: https://www.regjeringen.no/en/dokumenter/the-digital-norway-of-the-future/id3054645/?ch=2 [Accessed 2 July 2025].

Orwell, G. (2008 [1949]) *Nineteen Eighty-Four*. Penguin.
The World Happiness Report (2016, 2024) *Country Rankings* [Table]. Worldhappiness.report, [online], Available from: https://data.worldhappiness.report/table [Accessed 2 July 2025].
Utaker, R.G. (2025) 'De logget av sosiale medier og startet et tidsskrift: – Vi er ikke imot teknologi' ['They logged off social media and started a journal: – We are not against technology'], *Aftenposten*, [online] 8 June, Available from: https://www.aftenposten.no/kultur/i/gwn0lA/det-begynte-som-et-opproer-mot-meta-naa-er-det-blitt-en-liten-bevegelse-som-handler-om-analogt-velvaere [Accessed 2 July 2025].
White, M. (2025) *The White Lotus*. HBO Entertainment.

Preface

A.R.E. Taylor

The initial analogue idyll idea stemmed from my PhD fieldwork with prepper communities and infrastructure resilience planners, who often celebrated the affordances of analogue technologies over their digital equivalents. In early March 2020, I was awarded funding from the Culture-Media-Text Research Centre at the University of Winchester to organize a symposium dedicated to further exploring the analogue idyll concept. Later that month, the UK Government announced that the first national lockdown would come into force on 26 March to avoid the spread of COVID-19. On 16 April, that lockdown was further extended for three weeks. Conferences were being cancelled.

The analogue idyll symposium was meant to take place in person in July 2020 but it was clear that the event would need to be moved online if it was going to go ahead. By May 2020, the term 'Zoom fatigue' had entered popular discourse. We were all feeling exhausted from teaching online and from back-to-back Zoom meetings for both work and socializing. Among my friends and colleagues working in higher education, this sense of digital exhaustion was being exacerbated by feelings of FOMO ('fear of missing out') in response to the many online events that were now taking place. Teams and Zoom had made it easy and affordable to organize online seminars, book launches, and conferences, all of which were now suddenly multiplying. My pre-COVID academic calendar, which I had thought was busy, was nothing compared to what I was grappling with now. I considered postponing the analogue idyll symposium to avoid contributing to peoples' sense of digital exhaustion, including my own. It felt a little counterintuitive to run an event exploring nostalgia for the analogue via Zoom. As I thought about what I should do with the symposium, I began to experience a kind of Baader-Meinhof situation: I began seeing the analogue idyll everywhere. The topic felt more relevant than ever as nostalgic narratives reminiscing about pre-COVID 'analogue' life proliferated. People were missing the days when they could socialize offline, when they could hug and touch another human being, when work didn't revolve around videoconferencing platforms, when their phones weren't constantly pinging with COVID-19 exposure alerts. Locked inside our houses with little else to do but scroll through social media, binge-watch Netflix, or play online games with friends, many of us were longing for a pre-COVID past where life was not structured so heavily around digital media. Reliance on digital technologies during the pandemic seemed to be exacerbating concerns about the dangers of digital overuse, with these tools often positioned as exhausting, addictive, unnatural, unhealthy, and harmful. By contrast, the analogue – the offline, the outdoors, the physical touch of another human being – was celebrated as healthy and restorative, if sadly inaccessible for many.

I saw aspects of the analogue idyll surfacing in debates that were arising in the higher education sector between students and universities about online teaching vs in-person classroom-based ('analogue') teaching. Students were raising concerns that remote learning via Zoom was an inferior experience to learning on campus. Students were rightly concerned about value for money, given the exorbitant fees they pay in the UK. The debate over digital vs analogue teaching (not that these can be so easily separated in practice) tended to revolve around long-standing perceptions of the digital as *less than* the analogue, and therefore less value for money and less of a valuable experience. Remote learning was presented as less rich and rewarding than 'in-person' teaching. This reproduced long-standing tropes that celebrate offline (so-called 'face-to-face' or 'in-person') communication as superior to online interaction. Classroom-based 'analogue' teaching was being celebrated in ways that seemed to bolster notions that the offline is somehow more real, more authentic, more meaningful, and therefore worth paying for.

There is, of course, no reason why remote learning and teaching experiences should not be of the same quality or as rewarding as classroom-based teaching. Nevertheless, remote learning was persistently presented as inferior to in-person learning. This narrative was leveraged at different times by students and university management alike, to support their claims. Students felt shortchanged. Their analogue experience of university – the experience of being on campus, among friends, with face-to-face teaching, had suddenly been traded for a digital simulacrum. They were understandably struggling through online classes that in most cases had been designed for the seminar room or lecture theatre but had quickly been reconfigured for online delivery. Students wanted to be reimbursed. University management and recruitment teams, on the other hand, were presenting remote teaching as the continued delivery of a high-quality learning experience. As lockdowns eased, however, in response to fears that students may defer their degrees, many universities changed tune, highlighting the value of the analogue university experience and the importance of being on campus and in the classroom. In many cases this put staff and students at risk as universities desperately attempted to work at the maximum limit of social distancing, requiring complex hybrid delivery setups which often resulted in a poor learning experience equally for those offline and online. By this time, students had adjusted to online teaching. In-person attendance was dropping. Universities had to further promote the value of the analogue, if only to justify continuing to charge students rent for accommodation that many of them were not living in.

I felt the analogue idyll symposium could offer an opportunity to discuss these debates. While lockdown restrictions had been lifted (even though COVID-19 was still rapidly spreading), I decided to run the event online, albeit on the slightly later date of the 25 August 2020. The symposium was a success. It provided us with a critical (and, at times, cathartic) space to examine our own experiences of digital exhaustion during the pandemic and to engage more broadly with the emergence of 'the analogue' as a romanticized discourse that positions the pre-digital as more 'authentic' or 'real' than the digital. The symposium spawned not only this edited collection but also a 'digital exhaustion' research group that went on to publish the 2026 volume *Digital Exhaustion: Burnout, Fatigue, and Overload in the Age of Constant Connectivity*. Since 2020, concerns about the harms of digital culture have deepened. This book offers critical perspectives that situate these concerns within a broader cultural and political context, and ultimately suggests that the answers to 'digital' problems do not lie in simply retreating offline or longing for an idealized pre-internet past but in reimagining human-technology relations so that we may create more just, equitable, and enjoyable post-digital futures.

1

The Analogue Idyll: A New Myth for the Post-Digital Age?

A.R.E. Taylor

Introduction

Once upon a time, before the internet, before social media, and before smartphones, there existed an analogue world where people lived slower, simpler, and healthier lives. In this golden age, free from the addictions, distractions, and harms of digital culture, people were more in tune with nature. Their lives were less mediated, their personal relationships more authentic and fulfilling. People knew each other and talked to each other. They had *real* conversations face-to-face that were richer and more meaningful than the trivial and superficial interactions exchanged via social media or messaging apps. Without the constant notifications and pings of digital devices, people were less stressed and distracted, and could concentrate more deeply on the things that mattered. They spent their childhoods offline – outdoors in the 'real world' – not cooped up inside, endlessly scrolling on a glowing screen. They could identify trees, plants, and birds. They didn't have SatNavs. They knew how to map-read and embraced the possibility of getting lost, and the opportunity for discovery that enabled. People weren't tracked and profiled by big data corporations. They had privacy when they listened to their vinyl records, watched movies shot on *real* film with friends in the cinema, or purchased their weekly shopping with cash from their thriving local market or brick-and-mortar grocery store.

This is the analogue idyll: a compelling narrative that we increasingly encounter in contemporary culture that celebrates the virtues of analogue technologies and offline experiences. In popular expressions of the analogue idyll, digital technologies are presented as addictive, unnatural, unhealthy, and harmful. The analogue, by contrast, is presented as the remedy for our digital ailment.

If digital connectivity was once celebrated for its emancipatory potential (Castells, 1996), today it is being blamed for a proliferating array of societal ills (Turkle 2011, 2015; McChesney, 2013; van Dijck, 2013). Disinformation and fake news. Toxic viewpoints amplified by algorithmic filter bubbles. Social media and smartphone 'addiction'. Binge-watching and brain rot. Doom scrolling. Cyberbullying. Dwindling attention spans. Relentless data-based surveillance ('dataveillance'). Targeted political advertising. Internet scams and data breaches. Discourses about the negative impacts of constant connectivity on human health and wellbeing are proliferating in public life and the mass media. Social media platforms have sat at the centre of numerous moral panics (exacerbated by documentaries like *The Social Dilemma* [2020]), with some critics even suggesting that these platforms need 'warning labels' (Gerken, 2024). Digital discontent intensified during the COVID-19 pandemic, when many forms of social life were rerouted online, leading to a surge in utilization of digital technologies and backlash against them (Tai, Xiao, and He, 2021; Treré, 2021; Kuntsman, Martin, and Miyake, 2023). According to a 2025 survey carried out by the British Standards Institution (BSI, 2025), 47 per cent of young people aged between 16 and 21 would prefer to grow up in a world without internet. The survey also found that 68 per cent of respondents felt worse about themselves after spending time online. The rising tide of digital disillusionment was captured in 2018 when the word 'techlash' – a portmanteau of 'technology' and 'backlash', referring to a growing distrust of big tech companies – was shortlisted for the Oxford English Dictionary's 'Word of the Year'.

Amid this digital backlash there has been growing interest in – and celebration of – analogue media technologies and offline or 'IRL' ('in real life') experiences. Sales of vinyl records are the highest they have been in 30 years, with vinyl reportedly outselling streaming subscriptions (Titcomb, 2022) and independent record shops experiencing a major resurgence (Evitts and Trenholm, 2024) (Figure 1.1). Printed books are outselling eBooks, and Gen Z are apparently turning to libraries to escape the 'oversaturation and noise' of the digital world (Mac Donnell, 2024; see also Sweney, 2020). The jigsaw puzzle and boardgame industries are thriving (Sweney, 2021; Thorén, 2021; Kviat, 2024). Audio cassette production has been rejuvenated, and new cassette players are being brought to market (Ledsom, 2020; Savage, 2021). Polaroid has re-launched its line of instant film cameras, while Kodak is releasing its first new super 8 film camera in decades. Celluloid filmmaking is experiencing a resurgence, with an increasing number of films being shot on 35 mm and 70 mm (IMAX). Promotional campaigns for these productions often emphasize the immersive experience of watching the film in a cinema, aiming to draw

Figure 1.1: Riverman Records, an independent record store in Oxford (UK)

Source: A.R.E. Taylor

audiences away from digital streaming platforms and encourage them to head offline, back into movie theatres (the timespace of the cinema itself is being reframed by some theatre owners as a temporary 'digital detox' where audiences switch off their mobile phones and disconnect from the distractions of the digital world). Typewriter repair shops are opening. Postcards are making a comeback (Skopeliti, 2024). Artists are engaging with analogue printing and production methods, creating handcrafted zines, posters, and other paper-based ephemera (Crocket, 2024; Seedlings Studio, 2024). The renewed popularity of handmade goods and craft production has been linked to a broader cultural desire for 'the material, the tactile, the analogue' in a digital age (Luckman, 2013: 251). Sales of Moleskine notebooks are soaring, with 'good old fashioned' pen and paper increasingly perceived as a way to avoid smartphones and remain focused and creative (Yuan, 2021; Allen, 2024). Paper-based offices are making a comeback as workflow managers attribute productivity declines to employees' digital overload from continuous emails and notifications (Newport, 2021: 249; Taylor et al, 2026). According to one workplace productivity report, 96 per cent of officer workers in the United States prefer to work with hard copies over digital versions (Paper and Packaging, 2019). Paper is also being positioned as a solution to digital problems such as cyberattacks and data loss. Amid a growing number of large-scale cyberattacks disrupting the operations of major corporations, firms are being told that contingency plans should be backed up on paper in case they become inaccessible during a cyberattack (Tidy, 2025). Art projects like Kenneth Goldsmith's 'Printing out the Internet' (2013) and Michael Mandiberg's 'Print Wikipedia' (2015) underscore the fragility of digital storage media and suggest that, despite popular narratives that present digitization as technological 'progress', analogue formats like paper and microfilm remain better suited for long-term preservation. Even the protagonists in top-grossing blockbusters are celebrating the affordances of analogue media.[1]

This fascination with the analogue is further evidenced by the rising popularity of digital detox retreats and device-free public spaces, which promise to provide over-connected users with respite from the frenzied anxieties that come with 'always on' digital lifestyles (Fish, 2017; Sutton, 2017, 2020; Syvertsen and Enli, 2020, 2021; Syvertsen, 2020; Pennington, 2021). By creating offline spaces where the internet is not accessible or users are not allowed to use their digital devices, digital detox businesses strategically construct 'analogue' oases free from digital noise and distraction (Stäheli and Stoltenberg, 2024). Cafes, pubs, and night clubs are banning smartphones or establishing laptop-free and 'Wi-Fi-free' ('No-Fi') zones with the hope of rekindling vanishing forms of pre-digital sociality, and, in the process, conveniently reducing the number of laptop-using customers occupying tables for lengthy periods of time (Figure 1.2) (The Guardian 2024; Kviat, this volume). Samuel Smith's – a UK brewery which operates over 200 pubs – prohibits mobile phones, laptops, Kindles, TVs, and even electronic payments in their establishments. On their website they promote their pubs as 'havens

Figure 1.2: The Blue Bell is a cash-only pub in York (UK) and asks patrons not to converse on their mobile phones on the premises

Source: The Blue Bell, York

from the digital world' (Samuel Smith's, 2021). Celebrities are also cashing in on the digital detox trend, with footballer Cristiano Ronaldo opening a club in Madrid where mobile phones are banned (Monroe, 2025). In 2015, a photo of a sign in a London cafe that read 'We do not have WiFi … Talk to each other. Pretend it's 1995' went viral (Ewen, 2020). The sign struck a chord with many, inspiring other cafes to adopt similar positions. Lack of Wi-Fi (or restricted Wi-Fi) was reframed as an asset rather than a limitation – offering customers an opportunity to disconnect from digital pressures and reconnect with one another. The sign has since been widely reproduced, spawning multiple copies, with various versions now available for purchase online. The virality of this image is reflective of a wider nostalgia for forms of social interaction and spontaneous encounters that many imagine *used* to take place in cafes, as idealized in 1990s sitcoms like *Friends* (1994–2004), but have declined in the digital era, as customers sit isolated on their smartphones and laptops. As literary critic John Freeman (2009: 99) observes, 'Whereas once cafés were filled with people talking to one another or reading books or newspapers, now you will find people sitting alone before the glowing screen of their laptop, typing emails, working on documents'.

Popular modes of critiquing digital culture often express concern that social media and smartphones have eroded 'face-to-face' or 'in-person' conversation, resulting in superficial and inauthentic interactions.[2] For this reason, dating apps are in decline as people head offline in search of a 'more authentic way to find connections' (Macdonald, 2025). 'Logging off' clubs have also been established that run 'IRL events to reconnect with yourself and others' (loggingoffclub, 2025). Self-help books about 'digital exhaustion' (Leonardi, 2025) and 'digital minimalism' (Newport, 2019) aim to help over-connected users take control of the addictive design logics of social media platforms and smartphones. With a whole range of products and services now targeting digital disconnection, the analogue has become big business.

New markets have also emerged for 'dumb' digital devices that are not connected to the internet (in contrast to devices that are rendered 'smart' through internet access) (Bearne, 2022). In May 2024, Nokia launched a reconfigured version of their famous 3210 feature phone, which was originally released in 1999. The phone has no app functionality and an extremely low-spec camera, which, according to the press release, encourages users to 'return to simpler times' and 'get back to reality' (HMD, 2024a). Released by the phone manufacturer HMD (which stands for 'Human Mobile Devices'), the focus of the phone, we are told, is on 'conversations' rather than the 'likes and shares' of social media culture (HMD, 2024b). Even though the phone is technically a digital device, HMD nevertheless promote it as 'a solution for a digital detox in a world where almost four in ten (38%) of 16–24 years olds worry they spend too much time on their smartphone' (HMD, 2024a). Video games players are also heading offline, opting to play retro games without the stress that can come with online gaming (Banfield-Nwachi, 2025). Though not strictly analogue, the revival of these offline digital technologies is part of a larger economy dedicated to internet disconnection that has now surfaced.

This interest in analogue media and offline experiences has not gone unnoticed. Many journalists have now penned columns about the 'vinyl revival' or charted the 'return' of various analogue media technologies. Marketing strategists and trend analysts have developed insight papers about the turn to the offline in experience-orientated consumerism, which revolves around brick-and-mortar stores and immersive, multi-sensory experiences (GEP, 2024). In 2016, journalist David Sax mapped the rising popularity of all-things analogue in his book, *The Revenge of Analog: Real Things and Why They Matter*, which became an international bestseller. Media and cultural studies scholars have also explored the growing appeal of analogue technologies (Luckman, 2013; Cramer, 2014; Niemeyer, 2014; Wieghorst, 2021) and the different ways that users unplug, opt out, disengage, and abstain from using digital technologies or online services (Casemajor et al, 2015; Brennen, 2019; Kuntsman and Miyazaki, 2019; Natale and Treré, 2020; Fast, 2021; Helsper, 2021; Lomborg and Ytre-Arne, 2021; Syvertsen, 2022; Albris et al, 2024). Noting the prevalance of discourses and practices related to disengaging from digital culture, André Jansson and Paul C. Adams (2021: 3) have suggested that the study of digital disconnection will 'define social science in the 2020s', much in the same way that the study of networking dominated the 2000s and 2010s.[3] In short, despite the unending drive towards digitization that big tech platforms and governments around the world are encouraging, analogue technologies and offline products and experiences are thriving.

The analogue idyll

The chapters collected here explore this growing enthusiasm for the analogue across a diverse range of contexts. The conceptual framework of the 'analogue idyll' structures the volume and refers to discourses, representations, imaginaries, and practices that promote or celebrate the positive affordances of the analogue. In literary studies, an 'idyll' is a pastoral poem or prose composition that depicts a highly romanticized and idealized rural life, traditionally celebrating the simplicity and innocent pleasures of shepherding.[4] This idyllic life was often located nostalgically in a past 'golden age' that was imagined to be simpler and more meaningful, a world that has since been irrevocably lost.[5] In eighteenth-century Britain, in a society that was undergoing sweeping urbanization and industrialization, the English Romantics projected many of the values associated with the pastoral idyll onto the countryside. In romantic painting and literature, the countryside was often presented through a nostalgic lens as a slower-paced, healthier, and happier place – a remnant of the past – in contrast to the frenzy, vice, and artificiality of modern urban life. These romanticized and idealized representations of the countryside have been critically conceptualized as expressive of the 'rural idyll' (Strathern, 1982; Rapport and Overing, 2003). Historian Jeremy Burchardt (2017: 65) notes that, as a critical concept, the rural idyll 'is a useful shorthand to describe a congeries of positive attitudes towards, and representations of, rural life and landscapes'. More than simply demarcating a spatial division between the rural and urban, evocations or representations of the rural idyll often express broader concerns about temporality, technology, progress, modernity, and the future (Marx, 1964; Williams, 1973).

Many of the idealized values that have historically been accorded to the rural are now being attributed to the analogue. If the rural idyll offered a vision of a time and place free from the malaise of urban life, the analogue idyll offers a vision of a time and place free from the malaise of digital life. In contrast to the speed, artificiality, and toxicity of the digital world, the analogue world provides opportunities to slow down and experience more 'authentic' and 'meaningful' social interaction. The analogue is typically presented as an unmediated space that is closer to nature or more natural than the digital. As Clara Wieghorst and Lea Zierott (2021: 80) note, '"the analog" describes a pristine state or a space free of any media'. The analogue idyll presents an appealing possibility: that there are still spaces untouched by digital technology where one can escape the relentless demands of networked life.[6] Often this 'outside' is imagined quite literally as the outdoors, where one can immerse oneself in nature. The persistent association between the analogue world and the natural world may stem from the perception that analogue technologies are older than digital technologies and, as such, are part of an analogue past where human beings were 'more inherently connected to nature' (Sterne, 2016: 42).

Constructions of the analogue idyll often have a spatial dynamic that resembles that of the town/countryside divide that structures the rural idyll. With internet connectivity widely available in cities, efforts to disconnect from digital technologies and experience the analogue often involve heading out into the 'wild'. Rural and remote areas have thus become key sites for digital detox retreats, promising rest, rejuvenation, and purification from digital exhaustion (Figure 1.3).

As Vanessa Bartlett and Henrietta Bowden-Jones (2017: 11) note, 'Earth, food, and the natural environment are often seen as the ultimate antidote to digital overload'. The analogue idyll often abuts with popular imaginaries of 'wilderness' as a space far removed from the malaise of modernity (Cronon, 1996). Even if the internet and mobile phone coverage are increasingly accessible in wilderness areas, the wild is nevertheless imagined as an analogue space where one can momentarily escape digital life. In some

Figure 1.3: Camp Grounded, a digital detox retreat in the redwoods of North California (35mm analogue photo)

Source: Theodora Sutton

cases, the increasing reach of digital connectivity is also reconfiguring conceptualizations of 'wilderness'. As digital expansionism aggressively continues to accelerate (albeit unevenly, and in ways that often reproduce historical patterns of marginalization and inequality), Wi-Fi signals, mobile phone networks, and other forms of electromagnetic radiation (or 'e-smog') from telecommunications infrastructure increasingly 'pollute' wild spaces.[7] Apps like the White Spots app help people find signal-free 'dead zones' (sometimes known as 'white zones') with low levels of electromagnetic radiation (Pearce and Gretzel, 2012; Wieghorst and Zierott 2021: 83). We are now seeing conservation projects arising to protect the few remaining electromagnetic radiation-free zones where it is imagined that pristine disconnection can still be experienced. In France and Italy, two 'refuge zones' (in the province of Drôme in southeastern France and within the Vena del Gesso Regional Park in northeastern Italy) have been established by people who suffer from electromagnetic hypersensitivity (for whom the radiation emitted from digital devices negatively impacts their health and wellbeing). In the UK, protestors have challenged the installation of 4G masts in the Knoydart peninsula in the Scottish Highlands (Cocker, 2023), an area that has been deemed 'Britain's last wilderness' (Gillespie, 2016), partly due its lack of mobile phone and internet coverage. Digital divides certainly endure (Ali, 2021), but satellite-based internet projects (like Elon Musk's Starlink and Jeff Bezos' Project Kuiper) are driven by visions of ubiquitous connectivity, where consumers can access Amazon.com from locations as wild and remote as the Amazon rainforest. Jonathan Crary (2013: 30) observes that the reach of the internet means that 'no moment, place, or situation now exists in which one can *not* shop, consume, or exploit networked resources'. Indeed, the ultimate dream of surveillance capitalism is for internet access to be available 'everyware', with the eventual aim of eliminating spaces or times where it is possible to exist offline.

At the same time, the analogue idyll is not quite as spatially determined as the rural idyll. Disconnecting from the digital world does not always take place in rural areas or even outside – one can take time away from a digital device at home or within a city. It is not the argument of this book that digital connectivity has now thoroughly networked the rural, to the point that the rural idyll is no longer a viable myth and the analogue idyll has arisen to take its place. One of the reasons why digital detoxes often occur outside of urban environments is precisely because the rural is still perceived to be less networked than the urban. The emergence of new communications technologies – from the telegraph to the telephone to the internet – has always prompted concerns that the speed and stress of modernity will reach the countryside. In fact, the notion that the countryside is endangered or under threat is a central animating element of the rural idyll. Whether due to telecommunications, industrialization, developments in transportation, or urban sprawl and counter-urbanization, invocations of the rural idyll often position the countryside as existing in a state of imminent erasure. Similarly, evocations of the analogue idyll sometimes position the analogue as a domain that is in danger of vanishing as digital technologies and practices become ever more omnipresent. We see this in debates about the loss of handwriting skills as schools increasingly prioritize touch typing (Rosen, 2025), laments about the decline of print media (Freeman, 2009: 177–9), and concerns about the waning conversation skills of younger generations.[8]

The analogue has thus surfaced as an increasingly valuable repository of meaning in a sociocultural milieu that is overwhelmingly digital. In everyday parlance, it is not uncommon to speak of '*the* analogue', as a noun phrase, to refer broadly to the domain of the offline. In its noun construction, the analogue is typically evoked to describe an idealized space, practice, or way of life, that is perceived to be slower, and more authentic, natural, and real than the digital, which is construed as fake and artificial (Robinson, 2008: 21). This expansive understanding has allowed for a range of things beyond technology to be described as 'analogue', including reality itself (with the 'real' or 'analogue' world often opposed to the 'unreal' or 'unnatural' digital world), and, amid the rise of Artificial Intelligence (AI), the human. If someone prefers to take notes using a pen and paper rather than a smartphone, they may describe themselves as an 'analogue' person (Schrey, 2014: 33;

Table 1.1: Binary opposites that often align with popular conceptualizations of the analogue/digital distinction

Analogue	Digital
Non-computational	Computational
Offline	Online
Old	New
Natural	Artificial
Rural	Urban
Real	Fake
Material	Immaterial
Physical	Non-physical
Tactile	Non-tactile
Tangible	Intangible
True	False
Unmediated	Mediated
Healthy	Unhealthy

Taylor and Ewen, this volume). If one chooses to live offline, detached from the internet, one could be said to be living an 'analogue life' (Wieghorst and Zierott, 2021: 80). As Derek Robinson (2008: 21) summarizes, the term 'analogue' has come to connote 'something authentic and natural, against the artificial, arbitrarily truncated precision of the digital'.

Underpinning the analogue idyll are multiple binary opposites that often overlap and interlink (Table 1.1). The analogue is conceived as the opposite of digital, as old rather than new, as natural rather than unnatural. The next sections of this chapter explore some of these associations to more closely examine what is at stake in evocations of the analogue idyll.

Analogue, digital, post-digital

The term 'analogue' can span several meanings and senses. In the technical-scientific domain of audio-visual signal processing, analogue and digital refer to distinct signal formats. Analogue signals are continuously variable waves, while digital signals are transmitted in discrete units of countable binary 1s and 0s (Figure 1.4).[9] In some contexts, these underlying technical properties form the basis for the celebration of the analogue over the digital. For example, an audiophile may celebrate the sonic properties of vinyl records and express concern that 'something' is lost in the music when the continuous and variable waves of analogue audio are converted into the discrete binary units of digital data required for a compact disc (CD), MP3

Figure 1.4: A representation of analogue and digital signals

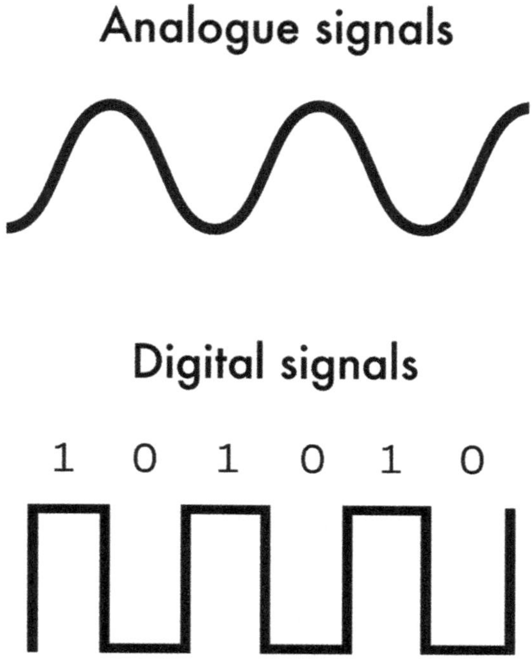

Source: A.R.E. Taylor

file, or streaming. As Pinch and Trocco (2002: 319) highlight, for some audio enthusiasts, 'digital sound is too perfect, too clean, too cold – they long instead for the imperfections of the warm, fuzzy, dirty analog sound'.[10]

Beyond the technical distinctions of information transmission, in common usage today, 'analogue' typically refers loosely to any technology that does not contain computational electronics and the term is often treated as a synonym for the 'non-computational', 'non-digital', or 'offline'. An example definition of this usage comes from journalist David Sax (2016: xiv), for whom the analogue,

> is the opposite of digital. Digital is the language of computers, the binary codes of 1's and 0's, which in endless combination allow computer hardware and software to communicate and calculate. If something is connected to the Internet, runs with the help of software, or is accessed by a computer, it is digital. Analog is the yin to digital's yang, the day to its night. It doesn't require a computer to function, and most often analog exists in the physical world (as opposed to the virtual one).

Media scholars may find this definition problematic because it positions the analogue and digital in a rigid binary opposition that overlooks historical and technical nuance. Distinctions between analogue and digital are complex and often overlap in practice. For example, digital technologies are manufactured from and rely on analogue materials, as visual culture theorist Florian Cramer (2014: 20) highlights: 'The electricity in a computer chip is analog, as its voltage can have arbitrary, undifferentiated values within a specific range … the sound waves produced by a sound card and a speaker are analog.' Another problem with using the term analogue to refer to anything non-computational or pre-computational is that this fails to capture historical complexity. Computers have not always been *digital* machines. Early electronic computers were known as analogue computers because they used continuously variable physical data (electrical, mechanical, or hydraulic quantities) to simulate models that stood in *analogical* relationship to the problem that was being modelled (Robinson, 2008). 'Analogue' in this sense refers to something that is comparable to something else (for example, when space engineers use Earth's deserts as 'analogue' sites to test equipment in environments or conditions similar – analogous – to Mars).[11] The operating of analogue computers was not based on binary digits (bits) but on the processing of continuous data from physical phenomena (such as voltage, temperature, water, or sound waves) that behave in ways analogous to the problem being solved. A mercury thermometer could be considered an example of a very basic analogue computer, where the expansion of the mercury is directly proportional to the temperature. Digital computers process data in discrete, binary units that can only be either 0 or 1, representing the 'on' and 'off' states of electrical switches within the computer, rather than a continuous range of values like an analogue computer.[12]

Just as the term analogue has come to stand in for everything and anything that exists 'outside computers' (Sterne, 2016: 37), in a colloquial sense 'digital' now often refers to anything computational or connected to the internet.[13] Originally stemming from the Latin *digitus*, meaning finger or toe (human 'digits'), digital broadly refers to 'something that is divided into discrete, countable units' (Cramer, 2014: 15). In this original sense of the term, many things that today we might not typically consider related to computing could be classed as digital. For example, the paper punch card system that employees once used to clock in and out of work is a digital system because it represents information (an employee's presence or absence) using discrete states in the form of holes punched in predefined positions that unambiguously log when an employee began and concluded their work. A guitar's fingerboard is technically a digital system because frets divide it into discrete notes, in contrast to the fingerboard of a violin, which is not divided by frets and therefore offers a continuous range of notes (Cramer, 2014: 16). An abacus could be considered an example of a non-electronic digital system because it represents values discretely via beads that are either at one end of the rod or the other (though it could also be considered an analogue system because the beads can slide across a continuous range of ambiguous positions on the rod).

Prior to the development of digital computing in the mid-twentieth century, analogue machines were the norm and, as such, the analogue nature of something was not remarkable.[14] It wasn't until the 1990s, when digital computers, mobile phones, and the internet had become pervasive, that the term 'analogue' became a meaningful category for understanding and describing anything that was 'not digital' (Sterne, 2016: 31; Palm, 2019: 644). In this sense, the analogue only became discenerable *after* the digital – which complicates simple media histories structured around the periodization of the 'analogue era' preceding the 'digital era'.

The analogue and digital have long co-existed. Scholars of media nostalgia have complicated simple distinctions that easily separate the analogue and the digital by exploring the vast range of practices through which analogue aesthetics are incorporated into digital media (Marks, 2002; Bishop, 2012; Niemeyer, 2014; Sapio, 2014; Schrey, 2014; Wieghorst, 2021). Social media filters enable users to add analogue-looking sepia tones, film grains, and effects to the photos they post online (Bartholeyns, 2014; see also Böhn, 2007). 'Analog Horror' is an internet-based subgenre of the found footage horror film that uses stock analogue effects, such as the static of cathode ray tube (CRT) TV sets or the colour shifts and visual distortions of VHS players, to make digital films feel more analogue (Balanzategui, 2023). In these cases, the analogue is not opposed to the digital but is enabled and remediated by digital technologies. Beyond digitized analogue aesthetics, much of contemporary analogue culture is enabled by the internet, which provides users with new opportunities to engage with the analogue. Artists post their analogue artworks or Polaroid photos on social media. Many analogue technologies – from vinyl records to discontinued Super 8 cameras – can only be found via online sites like eBay (see Palm, this volume). Digital detox retreats rely on websites and social media to attract customers (who, in turn, often post or blog about their revitalizing time offline when they return online). The supposed comeback of print magazines in the video gaming community has been enabled by cheap and accessible desktop publishing software (Stuart, 2022).

Analogue and digital technologies cannot be so easily disentangled from each other (Elwell, 2014; Thorén et al, 2019; Fortunati and O'Sullivan, 2020). As such, rather than imagine media histories in terms of linear trajectories from an 'analogue age' to a 'digital age', a close focus on technological complexity and overlaps highlights that we live in multiple 'ages' at the same time (Wieghorst, 2021: 208). Media scholars have developed concepts such as 'remediation' (Bolter and Grusin, 2000) and 'residual media' (Acland, 2007) to capture the ways that media technologies blend and blur into one another in ways that trouble simple histories constructed around rigid divisions between 'old' and 'new' media or 'analogue' and 'digital' technologies (Edgerton, 2006). In this sense, the term 'post-digital' has been used in an effort to develop a more heterochronic perspective on media history that 'allows for the simultaneity of the analogue and the digital' (Wieghorst, 2021: 222). The post-digital describes a cultural moment in which digital, analogue, and hybrid digital/analogue technologies, aesthetics, and experiences co-exist simultaneously (Cramer, 2014; Berry and Dieter, 2015). Similarly, trend analysts have used the term 'newstalgia' (Jackson, 2023) to refer to the incorporation of analogue or retro aesthetics in new digital devices, such as the music systems that are designed to appear like vintage radios and often feature a CD player, cassette player, analogue and digital radio, Bluetooth connectivity for streaming, and a turntable for playing vinyl records (Figure 1.5).

Figure 1.5: The GPO Chesterton music system combines multiple analogue and digital audio options and is nostalgically designed to look like a vintage music player

Source: GPO Retro

In a reversal of popular media histories, Alexander Galloway (2022) has argued that the golden age of analogue should not be understood as having occurred in a past, pre-digital era (when everything was analogue and therefore the analogue was unexceptional), but as unfolding *now*, in response to digitization (when the analogue has emerged as a distinct and revered category within a culture that is overwhelmingly digital). The golden age of analogue refers to our current fetishization and celebration of the analogue. Galloway argues that academia has also been experiencing its own golden age of analogue, highlighting that it is no coincidence that the 'distinctly nondigital themes' (Galloway, 2022: 211) of new materialism, affect, the senses, and object-oriented ontologies, all of which have arisen over the last two decades amidst the increasing digitization of society, as key areas of scholarly focus. These disciplinary turns to 'analogue' theorization mark an effort to place a renewed focus on the body and materiality in reaction to what is often understood as the dematerializing effects of digitization. Others have also complicated linear trajectories that place the analogue prior to the digital, noting that 'The analog only acquires its fascination by the juxtaposition with the digital' (Stäheli and Stoltenberg, 2024: 1061). It is only in response to the digital that the analogue came to be retrospectively understood as 'natural', 'real', or 'authentic'. The positive values attributed to the analogue today are thus retroactive (Sterne, 2016: 40).

There is thus considerable discussion and debate around current usage of the terms 'analogue' and 'digital' among media and communications scholars. While theorists may argue for the need to move beyond these simplistic binary concepts (Jurgenson, 2011; Thorén et al, 2019) or may propose hybrid concepts such as 'digilog' (Barry, 2014) or 'digitalized analogue' (Fortunati and O'Sullivan, 2020), in popular culture, imagined distinctions between the analogue and digital remain meaningful and are persistently evoked. Rather than engage in what Stäheli and Stoltenberg (2024: 1069) call the often 'fruitless conceptual discussion of defining analog and digital media', the framework of the analogue idyll directs attention to the practices, discourses, representations, and imaginaries through which positive values come to be attached to the analogue. In doing so, the contributors to this volume do not suggest that these positive values are inherent properties or fixed characteristics of the analogue technologies or experiences being discussed. Taking invocations of the analogue idyll seriously, the chapters collected here examine how different actors in different cultural contexts construct this narrative and how invocations of the analogue idyll makes distinctions between the analogue and digital socially meaningful. Whether or not we agree that the analogue is more authentic or less harmful than the digital, this is the broad cultural attitude and, as such, has an impact on how growing numbers of people live their lives and relate to digital technologies. As a critical concept the analogue idyll offers one way that we can begin to explore the socially constructed nature of narratives that romanticize or celebrate pre-digital and offline life.

By focusing on 'analogue' as much as 'disconnection' this volume expands the scope of work in digital disconnection studies by making room to consider cultural activities and forms of experience that might not so easily be described as 'disconnective practices' (Light, 2014) but nevertheless celebrate the non-digital. Indeed, popular esteem for the analogue is not only enacted through disconnection from digital media. While some engagements with analogue media may be carried out in response to a dislike for digital media, or to avoid the internet or abstain from using smartphones, enthusiasm for the analogue does not preclude engaging with digital technologies (one can simultaneously appreciate the sonic properties of vinyl records whilst still enjoying the convenience of streaming via Spotify).

Commodifying the analogue

A growing body of interdisciplinary scholarship is now analysing the many ways that individuals are disconnecting from the internet or spending time away from digital devices. Concerned about digital overuse, users are abandoning their social media accounts (Karppi, 2011, 2018; Portwood-Stacer, 2013; Foot 2014; Jorge, 2019; Baym, Wagman, and Persaud, 2020; Chia, Jorge, and Karppi, 2021; Feldman, 2021) and abstaining from using their smartphones and other technologies (Ytre-Arne et al, 2020). The scholarly field of digital disconnection studies builds on a longstanding body of media and communications research on media abstention, technology non-use, and news avoidance (Kraybill and Olshan, 1994; Selwyn, 2006; Portwood-Stacer, 2012a, 2012b, 2012c; Syvertsen, 2017; Hardey and Atkinson, 2018; Skovsgaard and Andersen, 2020). Research on technology resistance and refusal has demonstrated that focusing on non-use can yield insights into the social dimensions of technology that are equally as valuable as those derived from studying its use (Wyatt, Thomas, and Terranova, 2002; Selwyn, 2003, 2006; Oudshoorn and Pinch, 2003; Baumer et al, 2014; Wyatt, 2014). Practices of digital disconnection resemble earlier forms of media and technological resistance that emerged in response to modern technologies such as the telephone (Kline, 2003), cinema (Grieveson, 2004), and television (Krcmar, 2009). This larger body of work on technology refusal highlights that acts of disconnecting or opting out from techno-capitalism are not new and unfold across a continuum of positions, ranging from the ideological resistance of anti-tech radicals who live in a state of permanent politicized disconnection from the conveniences of modernity (such as the Unabomber, Ted Kaczynski [Fleming, 2022]), to those who strive to mindfully balance their consumption of media services (Kitchin and Fraser, 2020). What is new with current forms of digital disconnection is the normalization of these practices today. Early studies of media refusal largely focused on individuals or communities that existed on the peripheries of society by virtue of their disconnection from media services. These were people that were often constructed as 'deviant' because their lifestyles subverted the norm of connectivity – whether that was connection to media services or electrical and gas grids (Vannini and Taggert, 2013, 2015). In our current moment, dominated by the perception that digital technology has become perniciously pervasive, disconnection is the 'new normal'.

Another novelty about current practices of disconnection is that they are often not framed as subversive or radical. Rather, users are actively encouraged (increasingly, by tech companies themselves) to take time away from their digital devices. This invites us to reflect on who, beyond the user, benefits from digital disconnection and in whose interest moral panics about the dangers of digital overuse operate. Driven by discourses of techno-anxiety, the current popularity of digital disconnection has given rise to lucrative new markets and commercial opportunities that revolve around the nostalgic romanticization and commodification of 'the analogue', in its broadest sense. Self-help gurus and digital detox providers perpetuate the analogue idyll trope, promoting the restorative benefits of disconnecting in order to profit from the health concerns of over-connected users. The analogue is constructed as the domain for self-actualization, where users will rediscover their more authentic selves, regain control over their digital vices, and learn to become newly productive. In response to intensifying public and political fears about the impact of social media and smartphones on health and wellbeing, and in efforts to forestall government regulation, big tech companies have developed numerous apps and smartphone features (such as night modes and screen-time monitoring) that aim to promote 'digital wellbeing' (Mulvin, 2018; Beattie and Daubs, 2020). An entire economy of commercial apps has also arisen with the aim of helping users cultivate mindful connectivity by reminding them to take breaks to rest and repair (Kopitz, this volume). Many of these apps adopt the language of mindfulness and spirituality, encouraging users to subscribe to their services with promises of relaxation and recovery from digital overload. Mindfulness and other ancient practices formerly associated with Eastern philosophy now form the basis of what Ronald Purser (2019) calls the 'new capitalist spirituality'. Co-opted by the forces of commercialization and re-packaged as a 'privatized self-help technique' (Natale and Treré, 2020: 628), mindfulness has become a key tool of the digital disconnection industry and is often promoted on digital detox courses as a technique through which individuals can learn to more effectively manage their relationship with digital technologies. Efforts to resolve the many complex issues associated with digital technologies by simply disconnecting or taking a temporary 'detox' thus bolster an exploitative self-help industry which reconfigures social problems into personal problems where the individual user is blamed for their immoderate media consumption and charged with taking responsibility.

Digital wellbeing apps, self-help books, and digital detox retreats must be understood within the larger contexts of individual responsibilization and self-optimization that have been identified as characteristic features of neoliberal capitalism (Madsen, 2015; Syvertsen and Enli, 2020). Under neoliberal modes of power, governments step back from the provision of care and welfare. Rather than regulate big tech, responsibility is placed on individuals to exercise self-control and manage the pressures of constant connectivity by moderating their social media or smartphone use. This facilitates a culture in which individual users are chastized, blamed, or encouraged to feel ashamed if their media consumption is considered excessive (and, conversely, where individuals who practice disciplined moderation are praised and revered as good neoliberal subjects). Underpinning many digital wellbeing narratives is the ideology of 'productivity' (Gregg, 2018) and the idea that the self is a project that one is expected to continually work upon, to improve and optimize. For productivity gurus, while digital technologies can enhance productivity, one must carefully and mindfully navigate their addictive design, taking regular breaks or detoxes to avoid digital exhaustion (Taylor et al, 2026).

We must therefore exercise caution before interpreting analogue practices as forms of cultural resistance. Rather than challenge or dismantle the hegemony of digital connectivity, practices motivated by (or that reproduce) the analogue idyll can inadvertently serve to reinforce digital regimes without addressing the larger power structures of neoliberal capitalism, or the structural issues associated with hegemonic digital connectivity that led to the celebration of the analogue in the first place. For example, digital detoxing bolsters the dominance of digital culture by enabling exhausted users to take a temporary break so that they may return to the digital world, reinvigorated, refreshed, and ready to continue (best illustrated by the slogan 'Disconnect to Reconnect' which is used by the digital detox provider Camp Grounded). Detox retreats like Camp Grounded (located in the redwood forests of California and branded as a 'summer camp for adults' where mobile phones and digital devices are banned for the duration of their stay), provide users with a few carnivalesque days of relief from the digital world (Figure 1.6). Anthropological work on carnivals has argued that such events are licensed affairs permitted by dominant culture and, as such, provide a permissible rupture of hegemony. Anthropologist Max Gluckman asserted that while these 'rites of reversal obviously include a protest against the established order', they are ultimately 'intended to preserve and strengthen the established order' (Gluckman, 1965: 109). Rather than provide a space to imagine or realize new relational possibilities with digital technologies, digital detoxing tends to reinforce the existing social order by positioning temporary withdrawal as a simple fix.

Figure 1.6: Carnivalesque activities at the Camp Grounded digital detox retreat in North California (35mm analogue photo)

Source: Theodora Sutton

Analogue nostalgia

The commodification of the analogue today must also be seen in relation to the larger commodification of nostalgia in late capitalism. Nostalgia is a central feature of the analogue idyll. A number of media scholars have charted the rise of 'analogue nostalgia', exploring how the use of analogue media or aesthetics often mobilizes 'a longing for former times' (Wieghorst, 2021: 207; see also Marks, 2002; Bishop, 2012; Niemeyer, 2014; Sapio, 2014; Schrey, 2014; Wieghorst, 2021). Analogue nostalgia must be seen as part of fascination in contemporary culture not only with retro technologies – a trend that some have termed 'technostalgia' (Pinch and Reinecke, 2009; Bolin, 2015; van der Heijden, 2015) – but, more broadly with the retro and vintage across spheres of cultural and economic production, including fashion, design, and music (Reynolds, 2012; Guesdon and Le Guern, 2014; Lizardi, 2015). Nostalgia for the analogue today is not only evident in the major revival of interest in analogue media technologies but also in the markets that have emerged for new technologies that are designed to appear vintage.

Nostalgia has been widely theorized as a key cultural logic of the late twentieth and early twenty-first centuries (Davis, 1979; Jameson, 1988, 1991; Appadurai, 1996; Fisher, 2009, 2014). Svetlana Boym (2001: xiv) situates what she calls the 'global epidemic of nostalgia' in relation to the rise of 'cyberspace and the virtual global village', among other historical developments, and suggests that it is 'a defense mechanism in a time of accelerated rhythms of life'. Interpreted this way, current nostalgia for analogue media and experiences may be seen as the continuation of an extended reaction to the speed and frenzy associated with digital societies. For Fredric Jameson, postmodern culture's fixation with nostalgia was most evident in the proliferation of cinematic images that attempted to recreate the 'feel' or 'vibe' of past times. He introduced the concept of the 'retro mode' or 'nostalgia mode' (Jameson, 1988, 1991) to describe films that are less concerned with historical accuracy and more with imitating past aesthetic forms or offering stylized visions of the past based on selective, superficial, and often clichéd stereotypes associated with a historical period. He argued that this particular cinematic mode approaches 'the "past" through stylistic connotation, conveying "pastness" by the glossy qualities of the image' (Jameson, 1991: 19). Certainly, productive parallels can be drawn between Jameson's notion of the nostalgia film and the resurgence of analogue aesthetics and cultural forms today, where 'pastness' is conveyed by adding a film grain filter or sepia tint to an Instagram image. For Jameson and others (Fisher, 2009, 2014), the proliferation of nostalgia films pessimistically signalled a larger loss of our capacity to imagine or create new artistic modes. '[I]n a world in which stylistic innovation is no longer possible', he observed (1988: 18), 'all that is left is to imitate dead styles'.[15] Jameson's critique is itself infused with a certain sense of loss and nostalgia for a time of original film production, and overlooks the creative possibilities and subversive ways that nostalgic texts recreate, reconfigure, and recycle past aesthetics.

Evocations of the analogue idyll often express a yearning to return to simpler, safer, pre-digital times, and lament the loss of an enchanted and innocent analogue world. People from the 'bridge generation' (sometimes referred to as 'Xennials', a portmanteau of Generation X and Millennials), whose childhoods bridged the pre-internet and post-internet eras, may nostalgically remember analogue media from their youth, such as cassettes or VHS tapes. News articles in the popular press lament the generations 'who came of age online' and 'now feel deprived of real connections' because they 'grew up glued to their screens, and missed the joy of being human' (Hinsliff, 2025). For those born after the widespread adoption of the internet in the early 2000s, analogue nostalgia is often not related to direct experience but harks back to an imagined past that their parents or grandparents may have had. This is nicely captured by one of the top YouTube comments (with over 63,000 'thumb ups') on the music video for Lana Del Rey's 2011 song 'Video Games' – a video that makes heavy use of super-8 film throughout. The comment, posted in April 2020, states: 'She makes me feel nostalgic for things I haven't even experienced' (nitikasood1603, 2020). Feeling a sense of longing for a time that you have not directly lived through has been identified as a characteristic element of the experience of nostalgia in late modernity. Writing in the 1990s, cultural anthropologist Arjun Appadurai (1996) noted that marketing and merchandising techniques were increasingly in the business of manufacturing feelings of nostalgia that were not based on consumers' lived material realities. 'These forms of mass advertising', he highlighted (1996: 77), 'teach consumers to miss things they have never lost'. Appadurai employs a range of terms to try and capture this distinct mode of nostalgia, including 'imagined nostalgia' (77), 'armchair nostalgia' (78), and 'ersatz nostalgia' (82).

The reality crisis and the return of the real

In contemporary culture the analogue is often treated as a synonym for the 'real'. The offline, analogue world is typically conceptualized as 'real life' in contrast to the 'unreal' lives we construct and lead online. Digital detoxers take time away from social media in order to 'get back to reality'. Vinyl enthusiasts may claim that this music format sounds more 'real' than digital audio. Friendships that are developed offline 'in the real world' are often perceived to be more 'real' than the friends – or 'connections' – we make online.

The emergence of the analogue idyll in popular culture cannot be detached from the broader 'reality crisis' (Messeri, 2024) that has unfolded over the last two decades. The reality-distorting filter bubbles of social media algorithms, the 'alternative facts', 'fake news', and 'post-truth' politics that have accompanied the presidencies of Donald Trump, and the rise of Artificial Intelligence (AI), have all been blamed for loosening the fabric of reality. As philosopher Byung-Chul Han (2015: 43), suggests, 'Increasingly, virtualization and digitalization are making the real disappear'. Prior to these relatively recent reality-fracturing developments, since the 1980s cultural commentators have expressed concerns that new technologies and a highly simulated mediascape have given rise to a world in which it is increasingly difficult to distinguish the real from the fake.[16] The rapid circulation of mediatized images, the increasing affordability and popularity of cosmetic surgery, the enhanced verisimilitude of computer-generated simulations, the popularity of theming in urban planning and development (whereby cities and other spaces are modelled on theme parks, films, or other fantasy-worlds) and the market dominance of genetically modified foods that look and taste better than 'real' (natural/organic) foods, are just some of the developments that theorists argued were leading to an increasing blurriness between the 'real' and the 'fake' in everyday life (Eco, 1986; Sorkin, 1992; Baudrillard, 1994; Bryman,1999; Dear and Flusty, 1999; Haraway, 2000). Jean Baudrillard (1994) famously used the phrase 'hyperreality' to conceptualize the myriad ways that these developments were reconfiguring our relationship to reality.[17] Hyperreality referred to the ambiguous cultural terrain of late capitalism where differences between the fake and the real could no longer be discerned. The philosopher and semiotician Umberto Eco also used the phrase 'hyperreality' to describe the cultural condition whereby, even if the real and the fake could be identified, the fake was often preferred to the real (or appeared more 'real' than the real) (Eco, 1986; see also Best and Kellner, 1997). Eco famously contrasted his visit to Disneyland (California), which was full of animatronic animals, with a disappointing boat tour of the Mississippi River, where the 'real' crocodiles were not even visible. For Eco (1986: 45–46), Disneyland offered 'a fantasy more real than reality … and afterwards reality will always be inferior to it'.

The fetishization of the analogue today can be seen as part of a wider cultural thirst for the 'real' in a world that is increasingly assumed to be fake by default. 'Reality' has become a key selling point – or ingredient – in product marketing and advertising. We now live in a world where it is not unusual for burger chains to claim that their burgers are made from 'real' beef,

Figure 1.7: Reality is a key ingredient in McCain's potato smiles, which are 'made with real potato mash'

Source: A.R.E. Taylor

where potato chip packets assert that the chips within are made from 'real' potatoes, or ice cream labels highlight that their product contains 'real' cream (see Figure 1.7). Perhaps not unexpectedly, it is the big corporate brands, whose products are mass-produced and often assumed to contain artificial flavourings, that most often proclaim the 'realness' of their ingredients on the packaging of their products: the label on Heinz's tomato ketchup often states that their condiment is 'grown not made', the word 'real' dramatically dominates the label of Hellmann's mayonnaise; REAL Crisps is the name of a UK brand that manufacture hand-cooked potato chips.

We can draw parallels between these reality claims and the fixation with 'authenticity' in popular culture (Enli, 2015; Heřmanová, Skey, and Thurnell-Read, 2023). Amid anxieties about the stability of 'reality' and 'truth', authenticity has become a highly valued quality and a key buzzword across social media, tourism, politics, branding, and advertising. Politicians, celebrities, and products must now all stive to appear authentic to appeal to consumers' search for the real in an unreal world.

The association of the analogue with the real is partly underpinned by the perception that the analogue is physical, concrete, and tangible, in contrast to the digital, which is often understood as 'virtual', intangible, and existing in an electronic world behind the screens of our devices. The analogue is also often perceived as the domain of truth or truthfulness in contrast to the falsity of online sociality. The next two sections further explore these connections between the analogue and the real to trace how the 'digital' came to embody the falsity and 'de-realization' of contemporary societies.

The materiality of the digital

The persistent association of the analogue with the real is part of a larger history of representing digital media (and the internet in particular) as an immaterial or non-tactile domain that exists in an electronic or 'virtual' realm detached from material reality. This understanding is nicely illustrated by David Sax's (2016: xiv) previously cited claim that 'analog exists in the physical world (as opposed to the virtual one)'. Sax's definition reproduces an old but still prevalent binary view of the digital as a virtual world that is separate from the analogue or 'physical world'.[18] According to this view, the digital world is 'virtual', and the physical world is 'real' (Jurgenson, 2011). Popular metaphors like 'cyberspace', 'virtual world', 'information superhighway', and 'the cloud', which have been used to describe and imagine the internet and other electronic representations, have played an important role in structuring perceptions of digital computation as taking place in an ethereal 'elsewhere' removed from the 'real world' (Taylor, 2022: 216). William Gibson's 1984 novel *Neuromancer* popularized the term 'cyberspace', while films like *Tron* (1982) represented the software of a mainframe computer as an 8-bit electronic 'world' made up of neon-grids (Taylor, 2017: 46). This spatial representation of the digital as an electronic realm has also been established through popular films like *The Matrix* (1999), *Wreck-it-Ralph* (2012), and *The Emoji Movie* (2017). We can also see the trope of digital immateriality in Nicholas Negroponte's 1995 book *Being Digital*, in which he famously described digitization as a process involving a shift from 'atoms to bits' (Negroponte, 1995: 4). If atoms are understood to be the basic building blocks of physical matter, Negroponte presents digitization as a process of dematerialization, whereby material atoms are reconfigured into immaterial electronic bits in cyberspace.

While the metaphors that we use to apprehend the digital might appear mundane, political geographer Peter Vujakovic (1998: 158) highlights their social and political power: 'Common acceptance of particular metaphors (propagated or reinforced by the media) may lead to a limited view of an issue and the closure of constructive alternatives.' Metaphors that present the internet as a dematerialized, virtual space or world, overlook the materiality of the digital – or what computer scientist Paul Dourish (2017) has referred to as 'the stuff of bits' – and its social and environmental impacts. A significant body of work has now drawn attention to the materiality of the digital (Kirschenbaum, 2008; Blanchette, 2011; Horst and Miller, 2012; Reichert and Richterich, 2015). The binary digits ('bits') that make up the digital representations we see on screen are generated by physical computing machinery, which consists of various mineral and elemental components (Parikka, 2015; Starosielski, 2019), and an extensive infrastructure of fibre-optic cables, data centres, energy networks, human workers, and logistical arrangements of natural resources (Hogan, 2015; Parks and Starosielski, 2015; Starosielski, 2015; Rossiter, 2016; Plantin and Punathambekar, 2019; Taylor, 2019). To continue imagining the internet as an immaterial realm that is separate from material 'reality' is to overlook this massive infrastructure's geopolitical, social, and environmental entanglements with the 'real world' (Ortar et al, 2022).

The pervasive notion that the digital is a non-tactile and intangible realm removed from physical reality and human sensory experience has led to broader concerns that digital life marks a form of sensory impoverishment. Journalist Christine Rosen (2025: 77) in her book, *The Extinction of Experience: Reclaiming Our Humanity in a Digital World*, laments that 'As our world becomes ever more saturated with images and virtualizations, we shouldn't let our desire for technologies eclipse the human need to see, touch and make things with our hands.' For Rosen, the increasing digitization of everyday life robs us of the corporeal, sensory experience of material (analogue) things which she understands as essential to being human. Rosen suggests that the current popularity of handcrafted goods can be seen as a response to dissatisfaction with an increasingly virtual economy and reflective of a longing for sensory objects from which we are deprived in 'a world dominated by screens' (Rosen, 2025: 77).

Chris Anderson, the former editor-in-chief of the tech-focused magazine *WIRED*, has similarly blamed 'screens and personal computers for our lack of contact with physical objects' (Morozov, 2014). Discussing his involvement in a DIY 3D printing venture, Anderson reproduces the binary association of analogue physicality and digital non-physicality: 'digital natives are starting to hunger for life beyond the screen', he writes. 'Making something that starts virtual but quickly becomes tactile and usable in the everyday world is satisfying in a way that pure pixels are not' (Anderson, 2012: 18). Freeman (2009: 199–200) laments the decline of 'real-world meeting places' such as 'cafés, post offices, parks, cinemas, town centers', arguing that 'electronic communication leads us away from the physical world … We may rely heavily on the Internet, but we cannot touch it, taste it, or experience the indescribable feeling of togetherness that one gleans from face-to-face interaction'. Freeman ultimately calls for us to 'turn back to the real world and slow down' (2009: 203). The views of Freeman, Anderson, and Rosen are reflective of broader perceptions that the analogue is a physical and tacticle domain where we can engage in authentic sociality and hands-on sensory experiences. Anthropologists have advanced another perspective, arguing that we are no less human, authentic, or social when we are online – we just experience and engage with these differently (Horst and Miller, 2012). For many people, online interactions are just as rich and rewarding, or perhaps more so, than offline interactions. This popular association of the virtual with the non-tactile also overlooks the physical, sensorial, and haptic dimensions of the digital. Indeed, tactility and sensoriality – in the form of pings, vibrations, and other forms of sensory feedback – are in fact central to our interactions with digital devices, which are themselves physical, material objects (Paterson, 2007; Parisi, Paterson, and Archer, 2017; Parisi, 2018).

Analogue truth

The analogue (or the offline) is often presented as the domain of truth, where meaningful and authentic human relationships are developed and where one cannot deceptively 'hide' behind a screen. By contrast, the internet is presented as an artificial or false domain of filtered and staged photos, fake news, and superficial online 'friendships'. This distrust of the digital marks something of a shift from initial imaginaries of the internet. In the 1980s and 1990s, popular perceptions of the internet as a non-physical space detached from the material world led to celebrations of 'cyberspace' as an emancipatory realm for the presentation of one's 'true' self and the construction of authentic identity unfettered from the baggage of the material body. Removed from one's 'real-world' identity, those surfing the web could express their 'true' self without fear of judgement based on their personal appearance or life situation (Turkle, 1984, 1995, 1999). From this perspective, the process of digital dematerialization freed users from the materiality of the body and the social pressures associated with it (age, gender, ethnicity, class, and so on), enabling new forms of liberated subjectivity. As media scholar David Gauntlett (2000: 14) noted, because internet users 'cannot see each other, and are not obliged to reveal their real name or physical location, there is considerable scope for people to reveal secrets, discuss problems, or even enact, whole "identities" which they would never do in the real world'. This meant users were 'free to pick and choose genders, sexualities and personalities' (Gauntlett, 2000: 23), enabling some to feel they could 'express their "true" selves' when online (Gauntlett, 2000: 15). Early internet forums, message boards, and homepages were theorized as spaces for the performance of fluid, postmodern identities, where self-presentation could be 'achieved under optimal conditions' (Papacharissi, 2002: 644). 'In cyberspace', Mark Taylor and Esa Saarinen wrote (1994: 6), 'Identity becomes infinitely plastic in a play of images that knows no ends'. Users who played online games similarly celebrated the possibilities for truthful self-expression that their avatars afforded (Filiciak, 2003; Boellstorff, 2008; Nardi, 2010).

Of course, anxieties also emerged about the authenticity of these online presentations of self (Stone, 1991; Gauntlett, 2000: 14–15). Concerns about the ease with which users could create fake online identities to deceive others were reflected in films like *Hard Candy* (2005) and *Catfish* (2010), the plots of which revolve around individuals being manipulated online by users pretending to be someone they are not. In contrast to the liberating effect of early online forums where anonymity was the norm, the emergence of social media platforms placed an emphasis on 'pre-existing social relations' (boyd and Ellison, 2008: 221) and often revolved around photos of users, reconnecting identity and the body. However, rather than leading to more trust online, concerns with truthfulness now came to centre around users' strategic selecting and editing of photographs with the aim of creating the illusion of a perfect life, family, or body. Underpinning these concerns is an awareness of the ease with which digital technologies enable image manipulation.

Developments in imaging technologies since the 1990s, including Photoshop, computer generated imagery (CGI), and, more recently, social media filters and AI-generated imagery, have all given rise to concerns about the falsity of digital images. The relationship between digital images and reality has been an enduring topic for philosophers and media theorists since the 1980s. Digital images are understood to have an ontologically different relationship to reality than analogue images. Media theorist Lev Manovich (2001: 293), drawing on semiotic theory, refers to analogue cinema as the 'art of the index'. In semiotics (Peirce, 1931–48), an indexical sign implies the existence of an object in physical reality (for example, smoke is an index of fire, dark clouds are an index of rain, a footprint is an index of human presence). Analogue photos are indexes of the objects photographed by the camera: the material objects in the photo once existed in reality in order to be photographed (this is not to say that analogue images cannot be staged or manipulated). Analogue films thus 'largely consist of unmodified photographic recordings of real events which took place in real physical space' (Manovich, 2001: 294). The 'truth value' of analogue photography lay in its indexical relationship to reality.

Digital imaging technologies, however, can generate images that have no indexical relationship to any underlying reality. As Manovich (2001: 294) notes, with digital photography, 'everything can be simulated'. 'Rather than filming physical reality', Manovich (2001: 300) observes, digital technologies make it, 'possible to generate film-like scenes directly on a computer with the help of 3-D computer animation'. Digital technologies like CGI and AI-generated imagary can create images from scratch, without relying on re-presenting anything 'real' captured by a camera. For this reason, Michael Heim (1995: 70) argues that digitally produced visual entities are not representations: 'They do not re-present. They do not present again something that is already present somewhere else'. This is close to Baudrillard's (1994: 6) definition of digital simulation as a technique that 'is opposed to representation'. In his analyses of hyperreality, Baudrillard argued that digital modelling can generate the appearance of the 'real without origin or reality' (Baudrillard, 1994: 1). For this reason, he suggested, digital simulation 'threatens the difference between the "true" and the "false", the "real" and the "imaginary"' (Baudrillard, 1994: 3). Digital imagery's non-indexical relationship to reality means that it is often understood to lack truth value. Indeed, today, in a cultural climate where AI-generated videos, deep fakes, and filtered photos proliferate online, the digital is increasingly not trusted by default. Widespread cultural awareness of the potentially fabricated nature of digital imagery has led to dominant assumptions that most images posted on social media have been edited, processed by filters, or manipulated in some way. This perception has given rise to numerous social media trends based around efforts to introduce 'reality', 'truthfulness', and 'authenticity' online, such as the body positivity movement, 'de-influencing' trends (where 'influencing' has become synonymous with deception), and #NoFilter and Instagram Vs. Reality trends (whereby users share un-edited or 'truthful' selfies). Amid concerns about the spread of AI-generated imagery and the difficulty of distinguishing AI from reality, analogue nostalgia increasingly intersects with nostalgia for the human. While CGI once sat at the centre of conversations in the film industry about the replacement of analogue practices and human workers (such as special effects artists and stunt performers) by computerization, today, in the context of AI, it is the human element of traditional CGI that is now being foregrounded as a marker of analogue authenticity in an increasingly synthetic visual culture. CGI artists are now promoting human labour and skill as a celebrated element of digital production in contrast to AI-generated video production, in which humans are largely reduced to prompt inputters. As one 3D animator notes while showcasing his work on TikTok, 'like all my other videos, I've made this using the good old-fashioned 3D animation. You know, like the ones that you spend weeks creating' (Shortest Blockbusters, 2025). AI has provoked a longing for the perceived authenticity, materiality, and human element not only of analogue modes of image-making, but more broadly with pre-AI digital production processes. The values associated with the 'analogue' and 'digital' are thus constantly in flux and shift in relation to new developments in technology.

Chapter overview

The chapters in this volume trace a diverse range of responses to our current tech-saturated moment, investigating practices or products through which 'the analogue' is being made newly relevant as a domain of meaning, experience, and economic and imaginative investment. The 'analogue idyll' is presented as a framework for bringing together diverse and dynamic cultural expressions that promote the affordances of analogue technologies and offline experiences into a shared analytical space. The aim is not to debunk, demean, or dismiss expressions of the analogue idyll but to explore similarities, nuances, and differences across multifarious critiques of digital culture. As such, the chapters in this collection each show the rich range of ways that the analogue idyll is constructed or construed in different geographic, social, political, and cultural contexts.

The next chapter, authored by Michael Palm, focuses on what is perhaps the quintessential icon of analogue media: vinyl records. Vinyl has become a fetishized media format of the analogue economy. It is not only vinyl records themselves that are loaded with 'nostalgic charge' (Wieghorst, 2021: 209), but also their manufacturing process. In the UK, the popularity of vinyl has led to the opening of new vinyl pressing plants in the north of England. News coverage has not only celebrated the revival of vinyl but also the revival of factory work, nostalgically harking back to a golden era of 'analogue' labour that has declined with the rise of post-industrial, desk-bound, computer-based jobs. In the United States, the number of record processing plants has also increased dramatically over the last decade. In this chapter, Palm builds on his ethnographic research (2017; 2019) along vinyl's contemporary supply chains in the US to examine the contradictions of making and selling records in post-digital culture. Palm argues that vinyl record production is best seen as an example of 'mass craft', a term he uses to capture the ways in which these mass-produced media are invested with the authenticity typically associated with handmade craft goods. Moving beyond the analogue/digital binary, Palm's chapter introduces the concept of 'post-digital logistics', which he uses to theorize the symbiotic relationship between vinyl and the digital economy – where social media, streaming platforms, and online marketplaces, are all essential to vinyl's enduring popularity.

In Chapter 3, Alexandra Kviat examines efforts across the hospitality industry to rekindle forms of 'analogue' sociality. She explores the policies of growing numbers of cafes, pubs, and bars that ban or restrict Wi-Fi, laptop, or smartphone use on their premises. Informed by a survey of disconnection policies used by businesses operating in the United Kingdom, the United States, Canada, Australia, Denmark, and France, this chapter challenges some popular assumptions about their possible effects. Situating this phenomenon in a longer history of anxieties about technology, sociability and the 'third place' (Oldenburg, 1989), Kviat shows how disconnection policies contribute to table turnover, customer experience, brand positioning and promotion, and support the non-economic ambitions of hospitality entrepreneurs while simultaneously reinforcing normative ideas about who and what third places are for. Neither an antidote to hyperconnectivity nor a cynical marketing plot, such initiatives reflect broader tensions around precarity, community, and neoliberal politics.

In Chapter 4, the focus shifts outdoors. Dave McLaughlin analyses experiences of hikers on the Appalachian Trail in the United States to explore how long-distance walking holidays reproduce (and reconfigure) the analogue idyll. The health and wellbeing benefits of walking have long been widely celebrated but have taken on heightened significance as concerns spread about the addictive nature of digital technologies. McLaughlin explores how long-distance walkers, particularly hikers of colour, queer hikers, and Native American hikers, engage with (rather than reject) digital devices, challenging popular representations and imaginaries of the Appalachian Trail as an analogue idyll. He shows that the presence of digital devices does not necessarily negatively impact walkers' enjoyment of the Trail. Indeed, McLaughlin interrogates claims that position walking as a useful means of regaining our natural, authentic, harmonious selves outside of and away from digital connectivity. Instead, he shows how the analogue and digital interlink, conceptualizing the trail as a 'post-digital' walking experience.

In Chapter 5, Paul O'Connor traces the rise of a new digital divide that has emerged between those who have the economic means to disengage from social media and those who do not. The concept of the digital divide has traditionally described the disparity between those with access to information and communication technologies and those without, a gap often shaped by socioeconomic, geographic, and infrastructural inequalities. O'Connor approaches this concept from another angle, exploring the inequalities within highly digitized societies that enable some privileged people to disconnect from digital technologies while others from lower socioeconomic backgrounds have to remain connected, often for their work. In the hyper-connected Global North where digital connectivity is an increasing imperative for day-to-day life, taking time away from social media is something only a privileged few can do. O'Connor develops the concept of 'digital distance' to capture the ways in which some social media influencers strive to temporarily take a break from these platforms. Yet these influencers make sure they digitally capture their retreat to the analogue (which often take place in luxury resorts), taking photos of themselves relaxing by the pool or walking through idyllic countryside, and posting these when they return online. The analogue and digital are thus entangled as these influencers continue to use their digital technologies to document their time offline. For O'Connor, these acts of conspicuous digital detoxing valorize a specific form of luxury disconnection which is inaccessible to the majority of users.

In contrast to O'Connor's focus on social media abstention among influencers, Chapter 6 explores the theme of digital disconnection and privilege in relation to more traditional forms of celebrity. In this chapter, A.R.E. Taylor and Neil Ewen trace the emergence of the 'analogue celebrity'. They use this term to describe famous figures who engage in practices of digital disconnection that often hark back to earlier formations of pre-digital celebrity. These practices unfold across a spectrum of positions, from those who reportedly refuse to use email or mobile phones to those who take the occasional hiatus from social media. Taylor and Ewen focus on social media abstention among established celebrities. While social media has been widely theorized as a key tool for the practice of contemporary celebrity, Taylor and Ewen argue that the refusal of social media can add significant value to a celebrity's brand in the current cultural moment of digital backlash, providing celebrities with opportunities to align their images with ever-shifting constructions of authenticity, and to consolidate their fame within the attention economy.

The political economy of attention in the digital age is also the focus of Chapter 7. Malene Hornstrup Jespersen, Annika Isfeldt, and Kristoffer Albris explore the promotion of digital detox and disconnection courses in Denmark. This chapter is based on interviews with Danish companies that offer digital detoxing programmes. By combining analyses of these interviews with analyses of the companies' online marketing materials, Jespersen, Isfeldt, and Albris examine the different values behind these digital detox services, with some prioritizing mindfulness and others focusing on wilderness survival. They argue that what unites the companies, despite these different agendas, is the promise to their clients that taking a break from their digital lives will provide them with the means and tools to attain a more authentic and mindful presence in the world.

Further exploring the commercialization of mindfulness in the digital economy, in Chapter 8, Linda Kopitz analyses the booming market for mindfulness apps that aim to provide users with momentary respite from their busy digital lives. Kopitz analyses the advertisements of three leading mindfulness and meditation apps – *Calm*, *Headspace*, and *Meditopia* – that aim to enable users to momentarily re-connect with the analogue. Within the narratives of these advertisements, Kopitz identifies an inherent criticism of technologized, digital lifestyles, in contrast to the app experience, which is presented as natural, material, and analogue. Through this analysis, Kopitz thus draws our attention to the ways that representations of the analogue are mobilized by digital wellbeing companies and, by analysing these analogue-digital entanglements, explores the contradictions of using *digital* solutions to disconnect from *digital* problems.

Taken together these chapters demonstrate the rich range of practices and imaginaries through which the analogue is produced and reproduced within the idyll mode. In the examples and case studies presented in these chapters, the analogue is always valued highly, but what it is valued for, and why, differs significantly. By bringing together a range of representations and practices that celebrate the analogue under the framework of the 'analogue idyll' we do not suggest that all modes of engaging with the analogue are necessarily idealized or illusory, or romanticized. The concept of the rural idyll has been criticized for being 'far too dismissive' and assuming, a priori that positive views of the countryside are 'all idealised or even illusory, and that our task as academics is to debunk them' (Burchardt, 2017: 70). We certainly do not mean to engage in 'simplistic reductionism' (Burchardt, 2017: 72), where diverse and nuanced positive views of the analogue are lumped together as if they form a single cultural entity. Expressions of the analogue idyll necessarily unfold across a spectrum of positions and many people who prefer analogue technologies have good reasons for doing so. An interior designer may find that pen and paper are more conducive to capturing the free flow of thought than the notes app on their smartphone. An elderly person whose care alarm is connected to the analogue phone network may have well-founded fears about the 'digital switchover' to an internet-based phone system (Vallance, 2023). Someone undertaking a digital detox holiday in a rural B&B without broadband may be motivated by important mental health concerns. A nostalgic or romantic sentiment may shape many celebrations of the analogue but this should not prohibit us from taking seriously the criticism of – or scepticism towards – the digital that is often being expressed. Evocations of the analogue idyll invite us to question hegemonic assumptions about technological 'progress', techno-positivist narratives about innovation, and big tech marketing strategies that encourage users to continuously consume the latest digital devices, as well as the environmental impacts of these consumption practices.

At the same time, it is important to highlight the dangers of uncritically celebrating the analogue and to examine the privileged positioning from which evocations of the analogue idyll can emerge. Those who desire to disconnect or disengage from digital culture often reside in parts of the world that have high-speed and reliable broadband infrastructure. For those who live in areas with limited internet access, or who cannot afford digital devices, the analogue can constitute an involuntary and constraining daily reality. Even within countries with high levels of digital infrastructure, images and imaginaries that perpetuate the analogue idyll trope can reflect a privileged position that is detached from the socioeconomic realities of many people (the majority of whom simply can't afford to disconnect from digital technologies because their daily lives depend on these). There are also significant economic barriers that must be overcome in order to engage in certain analogue media cultures (vinyl, after all, is a lot more expensive than a streaming subscription). Practices of rejecting digital technology entirely, perhaps by living off grid to escape digital surveillance or lead a healthier life, do not help to reimagine digital culture differently or build a more just and equitable digital society, especially for those who can't afford to take refuge in the analogue. Rather than opt out entirely, it is through critical engagement with digital technologies and by placing political pressure on big tech companies and governments, that we can steer ourselves towards more equitable digital futures that are not based on unchecked data surveillance, burnout, and unsustainable upgrading (Cramer, 2014: 12; Casemajor et al, 2015: 863; Natale and Treré, 2020).

Notes

1 In the 2022 'whodunit' film *Glass Onion: A Knives Out Mystery*, tech billionaire Miles Bron (played by Edward Norton) doesn't own a mobile phone and insists on using an old fax machine, exclaiming that 'there's something about analogue'.

2 Curiously, the terms 'face-to-face' and 'in-person' are now often used to describe offline interactions, when it would be more accurate to simply use the term 'offline'. When we meet online via a video call, we are still technically conversing face-to-face (hence the name of Apple's video calling app 'FaceTime') and we are still technically meeting 'in-person' (can we meet someone 'out-of-person'?). The common use of 'face-to-face' and 'in-person' to describe offline meetings is a good example of how everyday language constructs the digital as disembodied and dematerialized.

3 For an analysis of the turn to disconnection in the social sciences beyond the study of digital culture, see Candea et al (2015).

4 The Greek poet Theocritus (mid-third century BC), whose poems were known as Idylls ('eidullia'), is often celebrated as the founder of Ancient Greek bucolic or pastoral poetry (Hunter, 2002).

5 Hesiod, writing in ninth century BC also looked back to a bucolic golden age (Short, 2006). The Roman poet Virgil would later further popularize the notion of the bucolic golden age, with his mythologizing of the unspoilt landscape of the Greek province of Arcadia in the *Eclogues* (Burchardt, 2002).

6 Cultural theorist Steven Shaviro (2003: 64) has observed: 'In an increasingly networked world, escape is nearly impossible. ... You cannot opt out of the network entirely, but at the very least, you can try to be connected a little less'. Others have also spoken of the difficulty of opting out of networked society when connectivity is made so easy. Galloway and Thacker (2007: 126) point out that with Wi-Fi, 'the very air you breathe is a domain of access'.

7 As Freeman (2009: 204) notes, 'Fewer and fewer parts of the world will be untouched by wireless Internet access'.

8 In *Conversation: A History of a Declining Art* (2006), Stephen Miller refers to mobile phones, iPods, computers, video games, and other digital technologies as 'conversation avoidance devices' (Miller, 2006: 282) and argues that these technologies are contributing to what he identifies as a decline in 'real' conversation, with people losing the skills and confidence to communicate offline. For similar perspectives, see Locke (1998) and Turkle (2015).

9 As Stewart Brand (1987: 18) notes, 'Analog is continuous, digital is discrete' (see also Goodman, 1968: 159–164).

10 On analogue purity, see Palm (this volume). For a discussion of analogue imperfection, see Wieghorst (2021: 211–212).

[11] This meaning of analogue to signify similarity or comparison stems from the Ancient Greek ἀνάλογος (análogos), meaning 'proportionate' (Galloway, 2022: 212).

[12] Derek Robinson (2008: 23) notes that 'digital computers operate on binary 1s and 0s rather than continuous voltages'.

[13] In fact, just as 'digital' now almost strictly refers to the computational, in everyday parlance the word 'technology' has increasingly come to refer narrowly to digital or computational hardware, a shift in meaning that overlooks entire histories of non-computational technology, from wooden spears to spectacles.

[14] This is not to say that there were not hierarchies attached to analogue technologies. For example, typewriters which are revered for their analogue charm today were once considered examples of the mechanization and industrialization of the writing process because they reduced the infinite variation, expressiveness, and continuous flow of cursive handwriting (an analogue system) into invariable, discrete letters (for this reason, in a technical sense, typewriters can be considered a digital writing system [see Cramer, 2014: 16]).

[15] Mark Fisher (2009: 59), citing Jameson's work on the postmodern nostalgia mode, similarly described contemporary culture as 'excessively nostalgic, given over to retrospection, incapable of generating any authentic novelty.'

[16] Concerns about technology replacing the real with something fake or unnatural have been a key pillar of modern social theory (see Jurgenson, 2013).

[17] Baudrillard's work presumes a singular and shared 'reality' or 'truth'. Such a perspective is decidedly western and overlooks the co-existence of multiple realities and versions of truth (Escobar, 2020; Messeri, 2024).

[18] The title of Sax's book itself, *The Revenge of Analog: Real Things and Why They Matter*, positions analogue technologies and experiences as 'real' things, and, by extension, suggests that digital technologies and experiences are somehow not real, or less real.

References

Acland, C.R. (2007) *Residual Media*. University of Minnesota Press.

Albris, K., Fast, K., Karlsen, F., Kaun, A., Lomborg, S., and Syvertsen, T. (eds) (2024) *The Digital Backlash and Paradoxes of Disconnection*. Nordicom.

Ali, C. (2021) *Farm Fresh Broadband: The Politics of Rural Connectivity*. MIT Press.

Allen, R. (2024) 'Moleskine Mania: How a Notebook Conquered the Digital Era', *The Walrus*, [online] 31 August, Available from: https://thewalrus.ca/moleskine/ [Accessed 2 November 2024].

Anderson, C. (2012) *Makers: The New Industrial Revolution*. Crown.

Appadurai, A. (1996) *Modernity at Large: Cultural Dimensions of Globalization*. University of Minnesota Press, pp 66–85.

Balanzategui, J. (2023) '"Shudder" and the Aesthetics and Platform Logics of Genre-Specific SVOD services', *Television & New Media*, 24(2): 156–172.

Banfield-Nwachi, M. (2025) '"There's no stress": gamers go offline in retro console revival', *The Guardian*, [online] 15 February, Available from: https://www.theguardian.com/games/2025/feb/15/theres-no-stress-gamers-go-offline-in-retro-console-revival [Accessed 15 February 2025].

Barry, E. (2014) 'Digilog Culture and the Vinyl Revival of the Early 21st Century', *American History Now: Explorations in Digital Curation*, [online] 30 April, Available from: https://americanhistorynow.org/index.html%3Fp=475.html [Accessed 12 January 2025].

Bartholeyns, G. (2014) 'The Instant Past: Nostalgia and Digital Retro Photography', in K. Niemeyer (ed) *Media and Nostalgia: Yearning for the Past, Present and Future*. Palgrave Macmillan, pp 51–69.

Bartlett, V. and Bowden-Jones, H. (eds) (2017) *Are We All Addicts Now? Digital Dependence*. Liverpool University Press.

Baudrillard, J. (1994) *Simulacra and Simulation*. Trans. S.F. Glaser. University of Michigan Press.

Baumer, E.P.S., Ames, M.G., Brubaker, J.R., Burrell, J., and Dourish, P. (2014) 'Refusing, Limiting, Departing: Why We Should Study Technology Non-use', *CHI EA '14: CHI '14 Extended Abstracts on Human Factors in Computing Systems*, pp 65–68.

Baym, N.K., Wagman, K.B., and Persaud, C.J. (2020) 'Mindfully Scrolling: Rethinking Facebook After Time Deactivated', *Social Media + Society*, 6(2): 1–10.

Bearne, S. (2022) 'Not Smart but Clever? The Return of "Dumphones"', *BBC News*, [online] 21 March, Available from: https://www.bbc.com/news/business-60763168 [Accessed 3 July 2023].

Beattie, A. and Daubs, M.S. (2020) 'Framing Digital Well-being as a Social Good', *First Monday*, 25(12), Available from: https://firstmonday.org/ojs/index.php/fm/article/view/10430/10012 [Accessed 5 May 2022].

Berry, D., and Dieter, M. (eds) (2015) *Post-digital Aesthetics: Art, Computation, and Design*. Palgrave Macmillan.

Best, S. and Kellner, D. (1997) *The Postmodern Turn*. Guilford Press.

Bishop, C. (2012) 'Digital Divide', *Artforum*, 51(1): 435–441.

Blanchette, J-F. (2011) 'A Material History of Bits', *Journal of the American Association for Information Science and Technology*, 62(6): 1042–1057.

Boellstorff, T. (2008) *Coming of Age in Second Life: An Anthropologist Explores the Virtually Human*. Princeton University Press.

Bolin, G. (2015) 'Passion and Nostalgia in Generational Media Experience', *European Journal of Cultural Studies*, 19(3): 250–264.

Bolter, J.D., and Grusin, R. (2000) *Remediation: Understanding New Media*. MIT Press.

boyd, d.m. and Ellison, N.B. (2008) 'Social Network Sites – Definition, History and Scholarship', *Journal of Computer-Mediated Communication*, 13: 210–230.

Boym, S. (2001) *The Future of Nostalgia*. Basic Books.

Brennen, B. (2019) *Opting Out of Digital Media*. Routledge.

Burchardt, J. (2002) *Paradise Lost: Rural Idyll and Social Change Since 1800*. I.B. Taurus.

Burchardt, J. (2017) 'The Rural Idyll: A Critique', in V. Elson and R. Shirley (eds) *Creating the Countryside: The Rural Idyll Past and Present.* Paul Holberton Publishing, pp 64–73.

Brand, S. (1987) *Media Lab: Inventing the Future at MIT.* Penguin.

Bryman, A. (1999) 'The Disneyization of Society', *The Sociological Review*, 47(1): 25–47.

BSI (2025) 'Half of Young People Want to Grow Up in a World Without Internet', *BSI Group*, [online] 20 May, Available from: https://www.bsigroup.com/en-GB/insights-and-media/media-centre/press-releases/2025/may/half-of-young-people-want-to-grow-up-in-a-world-without-internet/ [Accessed 20 May 2025].

Böhn, A. (2007) 'Nostalgia of the Media/in the Media', in W. Nöth and N. Bishara (eds) *Self-Reference in the Media.* de Gruyter, pp 143–153.

Candea, M., Cook, J., Trundle, C., and Yarrow, T. (eds) (2015) *Detachment: Essays on the Limits of Relational Thinking*, Manchester University Press.

Casemajor N., Couture, S., Delfin, M., Goerzen, M., and Delfanti, A. (2015) 'Non-participation in Digital Media: Toward a Framework of Mediated Political Action', *Media, Culture & Society*, 37(6): 850–866.

Castells, M. (1996) *The Rise of the Network Society.* Blackwells.

Chia, A., Jorge, A., and Karppi, T. (eds) (2021) *Social Media: Disconnection in the Age of the Techlash.* Rowman and Littlefield.

Cocker, A/R. (2023) 'The Remote Scottish Community of Knoydart Unites to Oppose Construction of Environmentally-damaging 4G Masts', *The Canary*, [online] 12 June, Available from: https://www.thecanary.co/uk/analysis/2023/06/12/the-remote-scottish-community-of-knoydart-unites-to-oppose-construction-of-environmentally-damaging-4g-masts/ [Accessed 3 July 2023].

Cramer, F. (2014) 'What is "Post-Digital"?' *APRJA*, 3(1): 10–24.

Crary, J. (2013) *24/7: Late Capitalism and the Ends of Sleep.* Verso.

Crocket, K. (2024) 'Analog Print in a Post-Digital World', *Creative Pinellas*, 1 May [online], Available from: https://creativepinellas.org/magazine/analog-print-in-a-post-digital-world/ [Accessed 14 February 2025].

Cronon, W. (1996), 'The Trouble with Wilderness: Or, Getting Back to the Wrong Nature', *Environmental History*, 1(1): 7–28.

Davis, F. (1979) *Yearning for Yesterday: A Sociology of Nostalgia.* Free Press.

Dear, M. and Flusty, S. (1999) 'The Postmodern Urban Condition', in M. Featherstone and S. Lash (eds) *Spaces of Culture: City, Nation, World.* Sage, pp 64–85.

Dourish, P. (2017) *The Stuff of Bits: An Essay on the Materialities of Information.* MIT Press.

Eco, U. (1986) *Travels in Hyperreality.* Trans. W. Weaver. Harcourt.

Edgerton, D. (2006) *The Shock of the Old: Technology and Global History Since 1900.* Profile Books.

Elwell J.S. (2014) 'The Transmediated Self: Life between the Digital and the Analog', *Convergence*, 20(2): 233–249.

Enli, G. (2015) *Mediated Authenticity: How the Media Constructs Reality*. Peter Lang.

Escobar, A. (2020) *Pluriversal Politics: The Real and the Possible*. Duke University Press.

Evitts, J. and Trenholm, M. (2024) 'Number of independent record shops in UK hits 10-year high', BBC News, [online] 6 March, Available from: https://www.bbc.com/news/newsbeat-68483084 [Accessed 8 April 2025].

Ewen, N. (2020) '"Talk to Each Other Like It's 1995": Mapping Nostalgia for the 1990s in Contemporary Media Culture', *Television & New Media*, 21(6): 574–580.

Fast, K. (2021) 'The Disconnection Turn: Three Facets of Disconnective Work in Post-Digital Capitalism', *Convergence*, 27(6): 1615–1630.

Feldman, Z. (2021) 'Quitting Digital Culture Rethinking Agency in a Beyond-Choice Ontology', in A. Chia, A. Jorge, and T. Karpi (eds) *Reckoning with Social Media: Disconnection in the Age of the Techlash*. Rowman & Littlefield, pp 103–130.

Filiciak, M. (2003) 'Hyperidentities: Postmodern Identity Patterns in Massively Multiplayer Online Role-Playing Games', in M.J.P. Wolf and B. Perron (eds) *The Videogame Theory Reader*. Routledge, pp 87–102.

Fish, A. (2017) 'Technology Retreats and the Politics of Social Media', *TripleC: Communication, Capitalism & Critique: Open Access Journal for a Global Sustainable Information Society*, 15(1): 355–369.

Fisher, M. (2014) *Ghosts of My Life Writings on Depression, Hauntology and Lost Futures*. Zer0 Books.

Fisher, M. (2009) *Capitalist Realism: Is There No Alternative?* Zer0 Books.

Fleming, S. (2022) 'The Unabomber and the Origins of Anti-Tech Radicalism', *Journal of Political Ideologies*, 27(2): 207–225.

Foot, K. (2014) 'The Online Emergence of Pushback on Social Media in the United States: A Historical Discourse Analysis', *International Journal of Communication*, 8: 1313–1342.

Fortunati, L. and O'Sullivan, L. (2020) 'Convergence Crosscurrents: Analog in the Digital and Digital in the Analog', *The Information Society*, 36(3): 160–166.

Freeman, J. (2009) *The Tyranny of E-mail*. Scribner.

Galloway, A.R. (2022) 'Golden Age of Analog', *Critical Inquiry*, 48(2): 211–232.

Galloway, A.R. and Thacker, E. (2007) *The Exploit: A Theory of Networks*. University of Minnesota Press.

Gauntlett, D. (ed) (2000) *Web Studies: Re-wiring Media Studies for the Digital Age*. Arnold.

GEP (2024) 'Experiential Marketing: How Brands Are Creating Immersive Consumer Experiences', *GEP*, [online] 16 November, Available from: https://www.gep.com/blog/mind/experiential-marketing-key-elements-for-success [Accessed 15 January 2025].

Gerken, T. (2024) 'Surgeon General Calls for Warning Labels on Social Media Platforms', *BBC News*, [online] 17 June, Available from: https://www.bbc.co.uk/news/articles/c98821dn27lo [Accessed 17 June 2024].
Gibson, W. (1984) *Neuromancer*. Ace Books.
Gillespie, K. (2016) 'The Last Wilderness', *The Scots Magazine*, [online] 6 June, Available from: https://www.scotsmagazine.com/articles/scotlanders-in-knoydart/ [Accessed 12 January 2025].
Gluckman, M. (1965) *Custom and Conflict in Africa*. Basil Blackwell.
Goldsmith, K. (2013) 'Printing out the Internet', exhibition at LABOR Art Gallery, Mexico City, 26 July–30 August.
Goodman, N. (1968) *Languages of Art: An Approach to a Theory of Symbols*. The Bobbs-Merrill Company, Inc.
Gregg, M. (2018), *Counterproductive: Time Management in the Knowledge Economy*. Duke University Press.
Grieveson, L. (2004) *Policing Cinema: Movies and Censorship in Early-Twentieth-Century America*. University of California Press.
Guesdon, M. and Le Guern, P. (2014) 'Retromania: Crisis of the Progressive Ideal and Pop Music Spectrality', in K. Niemeyer (ed) *Media and Nostalgia: Yearning for the Past, Present and Future*. Palgrave Macmillan, pp 70–82.
Han, B-C. (2015) *The Burnout Society*. Stanford University Press.
Haraway, D. (2000) *Donna Haraway and GM Foods*. Icon Books.
Hardey, M., and Atkinson, R. (2018) 'Disconnected: Non-Users of Information Communication Technologies', *Sociological Research Online*, 23(3): 553–571.
Heim, M. (1995) 'The Design of Virtual Reality', in M. Featherstone and R. Burrows (eds) *Cyberspace/Cyberbodies/Cyberpunk: Cultures of Technological Embodiment*. Sage, pp 65–78.
Helsper, E. (2021) *The Digital Disconnect*. SAGE.
Heřmanová, M., Skey, M., and Thurnell-Read, T. (eds) (2023) *Cultures of Authenticity*. Emerald.
Hinsliff, G. (2025) 'It's the age of regret: gen Z grew up glued to their screens, and missed the joy of being human', *The Guardian*, [online] 7 March, Available from: https://www.theguardian.com/commentisfree/2025/mar/07/gen-z-screens-human-generation-online [Accessed 8 March 2025].
HMD (2024) 'Get back to reality with the Nokia 3210', *Human Mobile Devices*, [online] 7 May, Available from: https://www.hmd.com/en_int/press/nokia-3210-press-release [Accessed 10 May 2024].
Hogan, M. (2015) 'Data Flows and Water Woes: The Utah Data Center', *Big Data & Society*, July–December: 1–12.
Horst, H.A., and Miller, D. (eds) (2012) *Digital Anthropology*. Berg.
Hunter, R. (2002), 'Introduction', in *Theocritus Idylls*. Trans. A. Verity. Oxford University Press, pp vii–xx.

Jackson, J. (2023) 'Exploring Y2K Newstalgia and its impact on consumer trends', *Trend Bible*, [online] 11 September, Available from: https://www.trendbible.com/exploring-y2k-newstalgia-and-its-impact-on-consumer-trends/ [Accessed 3 January 2024].

Jameson, F. (1988) 'Postmodernism and Consumer Society', in E.A. Kaplan (ed) *Postmodernism and Its Discontents: Theories, Practices*. Verso, pp 111–125.

Jameson, F. (1991) *Postmodernism, or, the Cultural Logic of Late Capitalism*. Duke University Press.

Jansson, A. and Adams, P.C. (2021) *Disentangling: The Geographies of Digital Disconnection*. Oxford University Press.

Jorge, A. (2019) 'Social Media, Interrupted: Users Recounting Temporary Disconnection on Instagram', *Social Media + Society*, 5(4): 1–19.

Jurgenson, N. (2011) 'Digital Dualism versus Augmented Reality', *The Society Pages*, [online] 24 February, Available from: https://thesocietypages.org/cyborgology/2011/02/24/digital-dualism-versus-augmented-reality/ [Accessed 7 July 2025].

Jurgenson, N. (2013) 'The Disconnectionists', *The New Inquiry*, [online] 13 November, Available from: https://thenewinquiry.com/the-disconnectionists/ [Accessed 7 July 2025].

Karppi, T. (2011) 'Digital Suicide and the Biopolitics of Leaving Facebook', *Transformations*, 20: 1–18.

Karppi, T. (2018) *Disconnect: Facebook's Affective Bonds*. University of Minnesota Press.

Kirschenbaum, M.G. (2008) *Mechanisms: New Media and the Forensic Imagination*. MIT Press.

Kitchin, R. and Fraser, A. (2020) *Slow Computing: Why We Need Balanced Digital Lives*. Bristol University Press.

Kline, R. (2003) 'Resisting Consumer Technology in Rural America: The Telephone and Electrification', in N. Oudshoorn and T. Pinch (eds) *How Users Matter: The Co-construction of Users and Technology*. MIT Press, pp 51–79.

Kraybill, D.B. and Olshan, M.A. (eds). (1994) *The Amish Struggle with Modernity*. University Press of New England.

Krcmar, M. (2009) *Living Without the Screen: Causes and Consequences of Life Without Television*. Norton.

Kuntsman, A. and Miyake, E. (2019) 'The Paradox and Continuum of Digital Disengagement: Denaturalising Digital Sociality and Technological Connectivity', *Media, Culture & Society*, 41(6): 901–913.

Kuntsman, A., Martin, S., and Miyake, E. (2023) *Digital Disengagement: COVID-19, Digital Justice and the Politics of Refusal*. Bristol University Press.

Kviat, A. (2024) 'Playful infrastructures: Building Communities Through Social Board Gaming', in G. Evans (ed) *Social and Cultural Infrastructure for People and Policy*. The British Academy, pp 74–86.

Latour, B. (2005) *Reassembling the Social: An Introduction to Actor-Network-Theory*. Oxford University Press.

Ledsom, A. (2020) 'The Walkman and Cassette Tapes Are Making a Comeback', *Forbes*, [online] 19 January, Available from: https://www.forbes.com/sites/alexledsom/2020/01/19/the-walkman-and-cassette-tapes-are-making-a-comeback/?sh=2d42d1d950ad [Accessed 14 July 2024].

Leonardi, P. (2025) *Digital Exhaustion: Simple Rules for Reclaiming Your Life*. Ebury Edge.

Light, B. (2014) *Disconnecting with Social Networking Sites*. Palgrave Macmillan.

Lizardi, R. (2015) *Mediated Nostalgia: Individual Memory and Contemporary Mass Media*. Lexington Books.

Locke, J.L. (1998) *The De-Voicing of Society: Why We Don't Talk to Each Other Anymore*. Touchstone.

Loggingoffclub (2025) 'Logging Off Club', *Instagram*, [online] Available from: https://www.instagram.com/loggingoffclub?igsh=cHR4eG11bmJoamFm [Accessed 3 January 2025].

Lomborg, S. and Ytre-Arne, B. (2021) 'Advancing Digital Disconnection Research: Introduction to the Special Issue', *Convergence*, 27(6): 1529–1535.

Luckman, S. (2013) 'The Aura of the Analogue in a Digital Age: Women's Crafts, Creative Markets and Home-Based Labour After Etsy', *Cultural Studies Review*, 19(1): 249–270.

MacDonnell, C. (2024) '"Reading Is so Sexy": Gen Z Turns to Physical Books and Libraries', *The Guardian*, [online] 9 February, Available from: https://www.theguardian.com/books/2024/feb/09/reading-is-so-sexy-gen-z-turns-to-physical-books-and-libraries [Accessed 12 March 2024].

Macdonald, F. (2025) 'Dating Apps Could Be in Trouble – Here's What Might Take Their Place', *BBC News*, [online] 14 February, Available from: https://www.bbc.co.uk/news/articles/cg7zxgxdggjo [Accessed 14 February 2025].

Madsen, O.J. (2015) *Optimizing the Self: Social Representations of Self-Help*. Routledge.

Mandiberg, M. (2015) 'Print Wikipedia', From Aaaaa! to ZZZap! exhibition at Denny Gallery, New York City, 18 June–11 July.

Manovich, L. (2001) 'What is Cinema?' *The Language of New Media*. MIT Press, pp 286–333.

Marks, L.U. (2002) *Touch: Sensuous Theory and Multisensory Media*. University of Minnesota Press.

Marx, L. (1964) *The Machine in the Garden*. Oxford University Press.

Messeri, L. (2024) *In the Land of the Unreal: Virtual and Other Realities in Los Angeles*. Duke University Press.

McChesney, R.W. (2013) *Digital Disconnection: How Capitalism is Turning the Internet Against Democracy*. New Press.

Miller, S. (2006) *Conversation: A History of a Declining Art*. Yale University Press.

Monroe, T. (2025) 'Cristiano Ronaldo Opens Ultra-Exclusive Club in Madrid... and You Can't Bring Your Phone!', *MSN*, [online] 27 November, Available from: https://www.msn.com/en-my/news/other/cristiano-ronaldo-opens-ultra-exclusive-club-in-madrid-and-you-can-t-bring-your-phone/ar-AA1RgwPo [accessed 29 November 2025].

Morozov, E. (2014) 'Making It', *The New Yorker*, [online] 5 January, Available from: https://www.newyorker.com/magazine/2014/01/13/making-it-2 [Accessed 3 February 2025].

Mulvin, D. (2018) 'Media Prophylaxis: Night Modes and the Politics of Preventing Harm', *Information & Culture*, 53(2): 175–202.

Nardi, B. (2010) *My Life as a Night Elf Priest: An Anthropological Account of World of Warcraft*. University of Michigan Press.

Natale, S. and Treré, E. (2020) 'Vinyl Won't Save Us: Reframing Disconnection as Engagement', *Media, Culture & Society*, 42(4): 626–633.

Negroponte, N. (1995) *Being Digital*. Hodder & Stoughton.

Newport, C. (2019) *Digital Minimalism: Choosing a Focused Life in a Noisy World*. Penguin.

Newport, C. (2021) *A World Without Email: Reimagining Work in the Age of Overload*. Penguin.

Niemeyer, K. (ed) (2014) *Media and Nostalgia: Yearning for the Past, Present and Future*. Palgrave Macmillan.

nitikasood1603 (2020) 'Lana Del Rey – Video Games' [YouTube comment], *YouTube*, [online] 11 April, Available from: https://www.youtube.com/watch?v=cE6wxDqdOV0 [Accessed 23 July 2024].

Oldenburg, R. (1989) *The Great Good Place: Cafes, Coffee Shops, Community Centers, Beauty Parlors, General Stores, Bars, Hangouts, and How They Get You Through the Day*. Paragon House.

Ortar, N., Taylor, A.R.E., Brodie, P., Johnson, A., Marquet, C., Pollio, A., and Cirolia, L. (2022) 'Powering "Smart" Futures: Data Centers and the Energopolitics of Digitalisation', in S. Pink, S. Abram, N. Ortar and K. Waltorp (eds) *Energy Futures: Anthropocene Challenges, Emerging Technologies and Everyday Life*. De Gruyter, pp 125–167.

Oudshoorn, N. and Pinch, T. (eds) (2003) *How Users Matter: The Co-construction of Users and Technology*. MIT Press.

Palm, M. (2017) 'Analog Backlog: Pressing Records During the Vinyl Revival', *Journal of Popular Music Studies*, 29(4): 1–14.

Palm, M. (2019) 'Keeping what Real? Vinyl Records and the Future of Independent Culture', *Convergence*, 25(4): 643–656.

Papacharissi, Z. (2002) 'The Presentation of Self in Virtual Life: Characteristics of Personal Home Pages', *Journalism & Mass Communication Quarterly*, 79(3): 534–806.

Paper and Packaging (2019) 'Digital Overload is Costing Employers Valuable Productivity, Many Employees Look to Paper for Focus Amid Digital Deluge', *Paper and Packaging*, [online] 8 October, Available from: https://www.paperandpackaging.org/sites/default/files/2019-10/Work%20Productivity%20Report%20Press%20Release%20FINAL.pdf [Accessed 8 November 2023].

Parikka, J. (2015) *A Geology of Media*. University of Minnesota Press.

Parisi, D., Paterson, M., and Archer, J.A. (2017) 'Haptic Media Studies', *New Media & Society*, 19(10): 1513–1522.

Parisi, D. (2018) *Archaeologies of Touch: Interfacing with Haptics from Electricity to Computing*. University of Minnesota Press.

Parks, L. and Starosielski, N. (eds) (2015) *Critical Studies of Media Infrastructures* University of Illinois Press.

Paterson, M. (2007) *The Senses of Touch: Haptics, Affects and Technologies*. Berg.

Pearce, P.L. and Gretzel, U. (2012) 'Tourism in Technology Dead Zones: Documenting Experiential Dimensions', *International Journal of Tourism Sciences*, 12(2): 1–20.

Peirce, C.S. (1931–1948) *Collected Papers*. Harvard University Press.

Pennington, N. (2021) 'Quitting Social Media: A Qualitative Exploration of Communication Outcomes', *Qualitative Research Reports in Communication*, 22(1): 30–38.

Pinch, T. and Trocco, F. (2002) *Analog Days: The Invention and Impact of the Moog Synthesizer*. Harvard University Press.

Pinch, T. and Reinecke, D. (2009) 'Technostalgia: How Old Gear Lives on in New Music', in K. Bijsterveld and J. van Dijck J (eds) *Sound Souvenirs: Audio Technologies, Memory and Cultural Practices*. Amsterdam University Press, pp 152–168.

Plantin, J-C. and Punathambekar, A. (2019) 'Digital Media Infrastructures: Pipes, Platforms, and Politics', *Media, Culture & Society*, 41(2): 163–174.

Portwood-Stacer, L. (2012a) 'How We Talk About Media Refusal, Part I', *Flow*, [online] 29 July, Available from: http://www.flowjournal.org/2012/07/how-we-talk-about-media-refusal-part-1/ [Accessed 23 July 2024].

Portwood-Stacer, L. (2012b) 'How We Talk About Media Refusal, Part II', *Flow*, [online] 10 September, Available from: http://www.flowjournal.org/2012/09/media-refusal-part-2-asceticism/ [Accessed 23 July 2024].

Portwood-Stacer, L. (2012c) 'How We Talk About Media Refusal, Part III', *Flow*, [online] 14 October, Available from: http://www.flowjournal.org/2012/10/how-we-talk-about-media-refusal-part-3-aesthetics/ [Accessed 23 July 2024].

Portwood-Stacer, L. (2013) 'Media Refusal and Conspicuous Non-Consumption: The Performative and Political Dimensions of Facebook Abstention', *New Media & Society*, 15(7): 1041–1057.

Purser, R. (2019) *McMindfulness: How Mindfulness Became the New Capitalist Spirituality*. Repeater.

Rapport, N., and Overing, J. (2003) 'The Rural Idyll', in N. Rapport and J. Overing (eds) *Social and Cultural Anthropology: The Key Concepts.* Routledge, pp 315–321.

Reichert, R. and Richterich, A. (2015) 'Digital Materialism', *Digital Culture & Society*, 1(1): 5–17.

Reynolds, S. (2012) *Retromania: Pop Culture's Addiction to Its Own Past.* Faber and Faber.

Robinson, D. (2008) 'Analog', in M. Fuller (ed) *Software Studies: A Lexicon.* MIT Press, pp 21–31.

Rosen, C. (2025) *The Extinction of Experience: Reclaiming Our Humanity in a Digital World.* Bodley Head.

Rossiter, N. (2016) *Software, Infrastructure, Labour: A Media Theory of Logistical Nightmares.* Routledge.

Samuel Smith's (2021) 'The Perfect Pub', *Samuel Smith's Brewery*, [online], Available from: https://samuelsmithsbrewery.co.uk/pubs/ [Accessed 18 May 2025].

Sapio, G. (2014) 'Homesick for Aged Home Movies: Why Do We Shoot Contemporary Family Videos in Old-Fashioned Ways?', in K. Niemeyer (ed) *Media and Nostalgia: Yearning for the Past, Present and Future.* Palgrave Macmillan, pp 39–50.

Savage, M. (2021) 'Vinyl Sales Soared Again in 2021, Thanks to Abba', *BBC News*, [online] 29 December, Available from: https://www.bbc.com/news/entertainment-arts-59808695 [Accessed 18 March 2024].

Sax, D. (2016) *The Revenge of Analog: Real Things and Why They Matter.* PublicAffairs.

Schrey, D. (2014) 'Analogue Nostalgia and the Aesthetics of Digital Remediation', in K. Niemeyer (ed) *Media and Nostalgia: Yearning for the Past, Present, and Future.* Palgrave Macmillan, pp 27–38.

Seedlings Studio (2024) 'Call for Writers and Artists for Our Exciting New Magazine' *Tik Tok*, [online] 1 October, Available from: https://www.tiktok.com/@seedlings.studio/video/7322439305396243745 [Accessed 12 October 2024].

Selwyn, N. (2003) 'Apart from Technology: Understanding People's Non-use of Information and Communication Technologies in Everyday Life', *Technology in Society: An International Journal*, 25(1): 99–116.

Selwyn, N. (2006) 'Digital Division or Digital Decision? A Study of Non-Users and Low-Users of Computers', *Poetics*, 34(4–5): 273–292.

Shaviro, S. (2003) *Connected, or What It Means to Love in the Network Society.* University of Minnesota Press.

Short, B. (2006) 'Idyllic Ruralities', in P. Cloke, T. Marsden, and P. Mooney (eds) *Handbook of Rural Studies.* Sage, pp 133–148.

Shortest Blockbusters (2025) 'Beneath the Surface', *TikTok*, [online] 25 June, Available from: https://www.tiktok.com/@shortestblockbusters/video/7519689595143900434 [Accessed 28 June 2025].

Skopeliti, C. (2024) '"We Need to Go Places and Touch Things": The People Turning Away from Smartphones', *The Guardian*, [online] 12 June, Available from: https://www.theguardian.com/lifeandstyle/article/2024/jun/12/young-people-turning-away-from-phones-social-media-addiction [Accessed 24 June 2024].

Skovsgaard, M. and Andersen, K. (2020) 'Conceptualizing News Avoidance: Towards a Shared Understanding of Different Causes and Potential Solutions', *Journalism Studies*, 21(4): 459–476.

Sorkin, M. (1992) *Variations on a Theme Park: The New American City and the End of Public Space*. Hill and Wang.

Starosielski, N. (2015) *The Undersea Network*. Duke University Press.

Starosielski, N. (2019) 'The Elements of Media Studies', *Media+Environment*, 1(1): 1–6.

Sterne, J. (2016) 'Analog', in B. Peters (ed) *Digital Keywords: A Vocabulary of Information Society & Culture*. Princeton University Press, pp 31–44.

Stone, A.R. (1991) 'Will the Real Body Please Stand Up? Boundary Stories about Virtual Cultures', in M. Benedikt (ed) *Cyberspace: First Steps*. MIT Press, pp 81–118.

Strathern, M. (1982) 'The Village as an Idea', in A.P. Cohen (ed) *Belonging: Identity and Social Organisation in British Rural Cultures*. Manchester University Press.

Stuart, K. (2022) 'Paper View: The Return of Video Game Magazines' *The Guardian*, [online] 11 March, Available from: https://www.theguardian.com/games/2022/mar/11/paper-view-the-return-of-video-game-magazines [Accessed 14 March 2024].

Stäheli, U. and Stoltenberg, L. (2024) 'Digital Detox Tourism: Practices of Analogization', *New Media & Society*, 26(2): 1056–1073.

Sutton, T. (2017) 'Disconnect to Reconnect: The Food/Technology Metaphor in Digital Detoxing', *First Monday*, 22(6), Available from: https://firstmonday.org/ojs/index.php/fm/article/view/7561/6310 [Accessed 23 July 2024].

Sutton, T. (2020) 'Digital Harm and Addiction: An Anthropological View', *Anthropology Today*, 36(1): 17–22.

Sweney, M. (2020) 'Pandemic Drives ebook and Audiobook Sales by UK Publishers to All-time High', *The Guardian*, [online] 14 November, Available from: https://www.theguardian.com/books/2020/nov/14/pandemic-drives-ebook-and-audiobook-sales-by-uk-publishers-to-all-time-high-covid [Accessed 24 July 2024].

Sweney, M. (2021) 'Digital Detox and Post-pandemic Catch-ups Fuel Board Game Boom', *The Guardian*, [online] 15 November, Available from: https://www.theguardian.com/lifeandstyle/2021/nov/15/digital-detox-and-post-pandemic-catch-ups-fuel-board-game-boom [Accessed 20 July 2024].

Syvertsen T. (2017) *Media Resistance: Protest, Dislike, Abstention*. Palgrave Pivot.

Syvertsen, T. and Enli, G. (2020) 'Digital Detox: Media Resistance and the Promise of Authenticity', *Convergence*, 26(5–6): 1269–1283.

Syvertsen, T. and Enli, G. (2021) 'Disconnect to Reconnect! Self help to Regain an Authentic Sense of Space Through Digital Detoxing', in A. Jansson and P.C. Adams (eds) *Disentangling: The Geographies of Digital Disconnection*. Oxford University Press, pp 227–252.

Syvertsen, T. (2020) *Digital Detox: The Politics of Disconnecting*. Emerald Publishing.

Syvertsen, T. (2022) 'Framing Digital Disconnection: Problem Definitions, Values, and Actions Among Digital Detox Organisers', *Convergence*, 29(3): 658–674.

Tai, Z., Xiao, Y., and He, B. (2021) 'Locked Down Through Virtual Disconnect: Navigating Life by Staying On/Off the Health QR Code During COVID-19 in China', *Convergence*, 27(6): 1648–1662.

Taylor, A.R.E. (2017) 'The Technoaesthetics of Data Centre "White Space"', *Imaginations: Journal of Cross-Cultural Visual Studies*, 8(2): 42–54.

Taylor, A.R.E. (2019) 'The Data Centre as Technological Wilderness', *Culture Machine*, 18: 1–30.

Taylor, A.R.E. (2022) 'Cloudwork: Data Centre Labour and the Maintenance of Media Infrastructure', in E. Costa, P.G. Lange, N. Haynes, and J. Sinanan (eds) *The Routledge Companion to Media Anthropology*. Routledge, pp 213–228.

Taylor, A.R.E., Kopitz, L., Soncul, Y., and Kviat, A. (2026) *Digital Exhaustion: Burnout, Fatigue, and Overload in the Age of Constant Connectivity*. Intellect.

Taylor, M. and Saarinen, E. (1994) *Imagologies: Media Philosophy*. Routledge.

The Guardian (2024) 'Snap out of it: Manchester club joins growing trend to ban phone cameras', *The Guardian*, [online] 14 December, Available from: https://www.theguardian.com/music/2024/dec/14/snap-out-of-it-manchester-club-joins-growing-trend-to-ban-phone-cameras [Accessed 8 January 2025].

Thorén, C. (2021) 'Pen, Paper, Dice ... Screen? Digital Resistance in the Swedish Tabletop Role-playing Game Community', *Convergence*, 27(3): 727–745.

Thorén, C., Edenius, M., Lundström, J.E. and Kitzmann, A. (2019) 'The Hipster's Ddilemma: What Iis Analogue or Digital in the Post-digital Society?' *Convergence*, 25(2): 324–339.

Tidy, J. (2025) 'Cyber Attack Contingency Plans Should Be Put on Paper, Firms Told', *BBC News*, [online] 14 October, Available from: https://www.bbc.co.uk/news/articles/ced61xv967lo [Accessed 14 October 2025].

Titcomb, J. (2022) 'Vinyl Grows Faster Than Streaming as Millennials Embrace the Record Player', *The Telegraph*, [online] 4 January, Available from: https://www.telegraph.co.uk/business/2022/01/04/vinyl-grows-faster-streaming-millennials-embrace-record-player/ [Accessed 6 January 2024].

Treré, E. (2021) 'Intensification, Discovery and Abandonment: Unearthing Global Ecologies of Dis/connection in Pandemic Times. *Convergence*, 27(6): 1663–1677.

Turkle, S. (1984) *The Second Self: Computers and the Human Spirit*. Simon and Schuster.

Turkle, S. (1995) *Life on the Screen: Identity in the Age of the Internet.* Simon and Schuster.

Turkle, S. (1999) 'Cyberspace and Identity', *Contemporary Sociology*, 28(6): 643–648.

Turkle, S. (2011) *Alone Together: Why We Expect More from Technology and Less from Each Other.* Basic Books.

Turkle, S. (2015) *Reclaiming Conversation: The Power of Talk in a Digital Age.* Penguin.

Vallance, C. (2023) 'Digital Landline Switch Paused for Vulnerable People', *BBC News*, [online] 18 December, Available from: https://www.bbc.co.uk/news/technology-67750235 [Accessed 21 December 2023].

van Dijck, J. (2013) *The Culture of Connectivity: A Critical History of Social Media.* Oxford University Press.

van der Heijden, T. (2015) 'Technostalgia of the Present: From Technologies of Memory to a Memory of Technologies', *NECSUS: European Journal of Media Studies*, 4(2): 103–121.

Vannini, P. and Taggart, J. (2013) 'Voluntary Simplicity, Involuntary Complexities, and the Pull of Remove: The Radical Ruralities of Off-grid Lifestyles', *Environment and Planning A*, 45: 295–311.

Vannini, P. and Taggart, J. (2015) *Off the Grid: Re-assembling Domestic Life.* Routledge.

Vujakovic, P. (1998) 'Reading Between the Lines: Using News Media Materials for Geography', *Journal of Geography in Higher Education*, 2(1): 147–155.

Wieghorst, C. and Zierott, L.P. (2021) 'Glossary', in T. Karppi, U. Stäheli, C. Wieghorst and L.P. Zierott (eds) *Undoing Networks.* University of Minnesota Press, pp 79–113.

Wieghorst, C. (2021) 'Analogue Nostalgia: Examining Critiques of Social Media', in A. Chia, A. Jorge, and T. Karppi (eds) *Reckoning with Social Media: Disconnection in the Age of the Techlash.* Rowman and Littlefield, pp 207–226.

Williams, R. (1973) *The Country and the City.* Oxford University Press.

Wyatt, S., Thomas, G., and Terranova, T. (2002) 'They Came, They Surfed, They Went Back to the Beach: Conceptualising Use and Non-use of the Internet', in S. Woolgard (ed) *Virtual Society? Technology, Cyberbole, Reality.* Oxford University Press, pp 23–40.

Wyatt, S. (2014) 'Bringing Users and Non-users into Being Across Methods and Disciplines', Refusing, Limiting, Departing: A CHI 2014 Workshop Considering Why We Should Study Technology Non-use, Toronto, Canada, 26 April–1 May.

Ytre-Arne, B., Syvertsen, T., Moe, H., and Karlsen, F. (2020) 'Temporal Ambivalences in Smartphone Use: Conflicting Flows, Conflicting Responsibilities', *New Media & Society*, 22(9): 1715–1732.

Yuan, Y. (2021) '"Think on Paper, Share Online": Interrogating the Sense of Slowness and Disconnection in the Rise of *Shouzhang* in China', in A. Jansson and P.C. Adams (eds) *Disentangling: The Geographies of Digital Disconnection.* Oxford University Press, pp 189–224.

2

Analogue Labour: The Mass Craft and Post-Digital Logistics of Making and Selling Vinyl Records

Michael Palm

Introduction

Vinyl records are unique. Historically, vinyl is the only popular media format to reverse sales trends after being superseded by a supposed upgrade. Left for dead at the turn of the twenty-first century, sales of new records have increased for 17 consecutive years. In 2021, consumers in the UK spent more money on new vinyl than on games for Sony PlayStations (King, 2022). The following year, in the US as well as the UK, new records outsold CDs for the first time in more than three decades (Smith, 2023a, 2023b). The markets for used records, meanwhile, have expanded and accelerated exponentially, thanks to search engines, online sales platforms, and digital payment networks. The website Discogs is an open-source database and marketplace, like a combination of Wikipedia and eBay but exclusively for music (the name is shortened from 'discographies'). In 2021 Discogs brokered the sale of over 13 million records and, as of April 2024, lists over 50 million records for sale, well over half of which are used.[1] Film, board games, notebooks, and other analogue media garner attention and engagement and even love from new initiates as well as holdovers, but no analogue format can hold a candle to vinyl's staying power.

Vinyl records are not only unique in the annals of media history, but each and every individual record is also unique at a technical level. Even when millions of copies are made, no two are identical. Their manufacture is so complex, unpredictable, and uncontrollable that, unlike digital files, every copy of a record contains its own microscopic constellation of imperfections and idiosyncrasies. Making records is hard work, and the coordination required is as complex as the chemistry. The production process requires sustained collaboration among discrete sets of technicians with rarified skills. Several pieces of elaborate, intricate machinery are essential, as are several suppliers, such as the handful of companies who process the polymer we colloquially call 'vinyl', not to mention support services like the postal services and shipping firms that deliver it to pressing plants. Pressing capacity globally has nearly tripled in the past decade, with more than a dozen new plants opening in the US since 2015, and independent record makers are addressing long-standing bottlenecks and pressure points along the supply chain for new records, which have been intensified by the major music labels' re-embrace of vinyl (not to mention the pandemic).

In this chapter's first two sections I elaborate the concepts of 'mass craft' and 'post-digital logistics' to help capture the (seeming) contradictions of vinyl's manufacturing and trafficking today. The science and technology of vinyl manufacturing remain virtually the same as they were in 1960, and plants still need to press hundreds of copies before they can break even. But now, on the other side of digitization, every record pressed embodies the type of authenticity traditionally reserved for the fruits of handmade goods. The purportedly warmer sound that vinyl records exude from their grooves is culturally coded, but there are technological reasons for it. Digital media and craft work are not mutually exclusive, of course, but the labour of records' fabrication has acquired new appreciation for its pre-digital tools and processes, akin to the rearticulated pleasures that now flow from listening to vinyl. Making records may still technically qualify as mass production, but it has become shot through with the authenticity and uncertainty of craft work. Meanwhile, the concept of 'post-digital' was developed to describe aesthetic experiences, but it is applicable to commerce as well as culture and to the trafficking of used records as well the manufacture of new ones (Berry and Dieter, 2015; Mazierska, Gillon, and Rigg, 2019). Accordingly, I find it more accurate and instructive to consider vinyl today a post-digital medium, rather than strictly an analogue one, and this tripartite distinction applies to vinyl labour as well as listening. Vinyl's contemporary supply chains demonstrate how digital and non-digital media are used in tandem in order to organize the production and distribution of most any cultural commodities that might contribute to an idyllically analogue experience.

Furthermore, the notion of an analogue idyll is itself post-digital, only making sense now that digitization has become pervasive. Vinyl occupies a privileged position within our new analogue imaginary, and the format's revived economy demonstrates the usefulness of remembering that an analogue idyll can only exist as digital culture's antithesis, rather than romantically reversing the historical relationship between digital realities and analogue imaginaries. In the two sections that follow, I consider how an analogue idyll informs the experience of making and selling vinyl records, rather than listening to them, and then in the conclusion I consider how the contemporary work of making and selling records can inform our understanding and appreciation of some of labour's post-digital conditions. The original idyll idea was a reaction to industrialization, proffering an alternative that was spatially based in the rural. There was also an element of slowness, in contrast to the speed of factory production and the bustle of the city. Notions of an analogue idyll are still based on the temporal binary of fast and slow, even if the urban/rural distinction no longer holds. Instead, a second temporal horizon animates the analogue idyll, namely the pre-digital past. By characterizing the manufacturing and trafficking of vinyl records as post-digital labour, rather than analogue labour, the jobs along vinyl's supply chains come into focus as representative of work today, rather than as an idyllic exception.

The mass craft of making records

The fabrication of vinyl records entails several different kinds of work. The program for the first annual *Making Vinyl* conference, held in 2017, welcomed 'great minds all along the vinyl value chain – from the pressing plants to the mastering facilities and PVC manufacturers [polyvinyl chloride, records' raw material] to the plater/stamper makers and lathe cutters to the labels and indie retailers' (Jaffee, 2017). The various labour processes of vinyl production are distinct and far more different than similar; experientially, they share little other than contributing to the same final product. What they do share is an artisanal ethos and a craft sensibility. While researching a book about the contemporary vinyl economy and the labour sustaining it, I've met scores of people who work along the supply chain for new records, and the majority of them enjoy their jobs. Most are motivated, at least in part, by a love of music, and many of them specifically love records. But these loves are distinct from their passion for the work itself, the motivation to master the particular sets of skills required to move records from one point to the next along their convoluted journey from studio to stereo.

The fabrication of vinyl records entails three distinct steps: mastering, plating, and then pressing. Record presses stand at the end of a long production chain, coming after two preliminary operations, and it is understandable that pressing records receives more attention than the stations upstream. After all, a press is where a record finally appears. A blob of heated plastic, known as a puck or biscuit, is flattened between two ridged metal plates and presto – a record! But at that point, it's usually been weeks if not months (or, during the pandemic, even years) since the music pressed into a record's grooves was recorded.[2] Ideally the decision to release a song or album on vinyl was made before anyone enters the studio, because the same recording will sound different emanating from a record's grooves than it will on tape or transposed into binary code (not necessarily better, mind you, but different).

Mastering is the preparation and transfer of a studio mix to a medium specifically designed for distribution or duplication; in the case of vinyl, that medium is almost always lacquer discs. Music recorded in a professional studio used to be committed to reels of analogue tape, and a fraction still is, but now most mastering engineers expect to receive high-resolution digital files. For most music that will end up on vinyl, its first relocation – from the recording studio to the mastering studio – happens over the internet. Once a recording is transferred onto a lacquer disc, however, no amount of digital technology will suffice for moving it to the next stage, electroplating. (Furthermore, lacquers begin to degrade after forty-eight hours, and in order to avoid the added costs of expedited weekend shipping, mastering studios avoid cutting lacquers on Fridays.) Some recording studios handle mastering in-house, and a few can even master for vinyl, but studio recordings bound for any format are usually sent to another facility for mastering. And while there is no shortage of mastering studios out there, few are equipped to cut lacquers. The reason for this has less to do with the technical proficiency required to skillfully cut a lacquer, which is substantial (see Figure 2.1), than with a physical lack of functional or salvageable lathes left in the world. Furthermore, the lacquer discs themselves are only being manufactured by one company worldwide, based in Japan, and the specialized styluses used to cut lacquers are only made by two.[3] This dearth of supplies means that most mastering studios have them delivered from far away, and the paucity of available lathes worldwide means that these same studios in turn ship their mastered lacquers long distances as well.

The middle stage of vinyl's three-part production process is even more precarious than the first. Lacquer discs are used to fashion metal plates or 'stampers', outfitted with ridges that provide the contours of a particular record's grooves. Lacquer masters can be played on a turntable, but they're soft enough that they'll degrade each time, so they are reproduced in a more durable form for replication, in a procedure known as electroplating. Vinyl's revived popularity led to a backlog in pressing plants, which has largely been addressed by a spate of new openings in the last decade, but 'the real problem is not in the pressing – the bottleneck is in the electroplating' (Herman, 2015). A handful of mastering studios handle plating in-house, as do a few high-end pressing plants, but in the US, I only know of three dedicated facilities for electroplating: Mastercraft, Inc. in Elizabeth, New Jersey; Gotta Groove Records Electromastering Lab in Columbus, Ohio; and Symcon in Charlotte, NC. The chemical process of electroplating generates toxic waste and draws regular visits from federal Environmental Protection Agency

Figure 2.1: Bob Weston, Chicago Mastering Service, etching an identifier onto a mastered lacquer

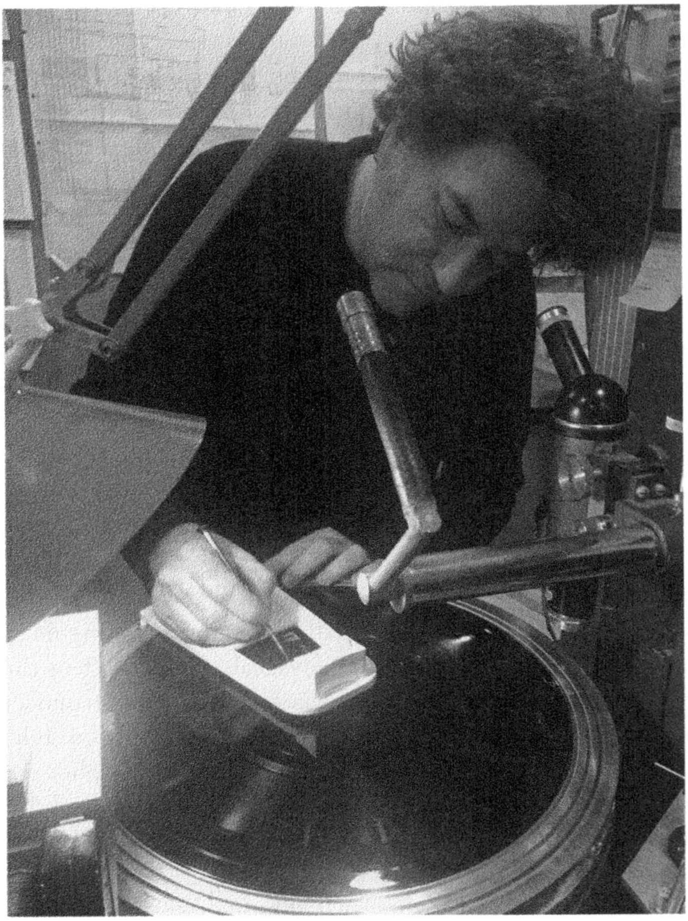

Source: Michael Palm

(EPA) inspectors. During a 2018 *Making Vinyl* session about new plants, several panellists stressed the importance of healthy relationships with local communities, and they all reported that their municipal governments had been supportive and encouraging. None of the new proprietors claimed to have considered plating in-house, however, and one explained that local officials tend to 'cringe' at even the mention of it, given the toxicity.

Once recorded music destined for vinyl has been transmogrified twice over, it's delivered to a pressing plant. Technically, people don't press records; machines do, and making records undeniably qualifies as mass production. The minimum run at most pressing plants today is several hundred (as opposed to several thousand during the format's heyday), and record pressing is massive in terms of the machinery as well as the scale. Vinyl is (too) often lionized for its purportedly warmer sound, but there is nothing inviting about the physics of fabricating records. A record press weighs thousands of pounds and is extremely loud when running. The mechanics are downright violent, a sort of controlled frenzy with thousands of pounds of pressure being hydraulically applied to heated plastic pucks that are fed into a press several times per minute. Pressing records is essentially factory work, and the human labour resembles that on an assembly line: loading the proper plates, monitoring the presses, and making necessary adjustments. At a record plant, however, these adjustments are never ending. Presses are 'finicky by nature', and even mid-size plants may employ 'a full-time machinist in house who repairs and makes replacement parts for the presses' (Eastes, undated). The process itself is not only volatile but surprisingly delicate – at one point during a *Making Vinyl* panel about starting up a pressing plant, dust was decried as 'the enemy'. Changes in humidity or barometric pressure can wreak havoc, and even seemingly harmless variables like 'a cool breeze can change a record', so a plant's temperature is continually monitored and tweaked, relying more on feel and experience than measurements. During the panel about starting up, one owner suggested hiring musicians to work the presses, since they tend not only to love music and often records specifically, but most also have experience solving problems that arise from relying on temperamental machinery for their work (Blocker, 2018). The clicks and pops and other imperfections of vinyl as a listening experience are often coded as charming, but for the workers along the vinyl supply chain for new records, the inevitability of imperfection turns their production into something approaching Sisyphean.

The three main stages of vinyl manufacturing – mastering, plating, pressing – take place in facilities that resemble, respectively, a studio, a laboratory, and a factory. All three entail: labour that is skilled and rarified; machinery that is expensive, intricate, and temperamental; and physics and chemistry that are precise yet often unpredictable. Each phase of vinyl production is a paragon of complexity unto itself, while the coordination between the points of vinyl's supply chain is arguably even knottier than any of the work done at a particular point along the way. And while the fabrication of new records may culminate at a pressing plant, their supply chain doesn't end there (after all, what good is a record if it never leaves the plant?). As I describe in the following section, the logistics of vinyl's contemporary supply chains involves not only shipping records from plants to distributors, labels, shops (where they are in turn shipped to customers), but also plants' procurement of key elements and supplies, such as record sleeves and custom-printed covers, bundled by the hundred or thousand. And there is, of course, the actual vinyl material itself, barrelled or boxed in a rainbow of colours, often after being shipped from another factory hundreds or even thousands of miles away.[4]

Among researchers of work there's a venerable tradition, from Marx on down, of conjoining new adjectives to the broad umbrella category of labour. I was no stranger to the practice when I first took an interest in the contemporary manufacture of vinyl records, during the mid-2010s, and I began referring to the process as 'analogue labour'. Not only does vinyl exude what Susan Luckman has called 'the aura of the analogue in the digital age', but the term 'digital labour' was having a moment among scholars bent on capturing what was distinguishing work now from before (Luckman, 2013).[5] In the same way that vinyl consumption is commonly differentiated from digital music experiences, I found it tempting to mark vinyl labour as an analogue exception to the new digital rules. But of course no tool belt today is complete without a smart phone, and digital technology helps connect workers along the supply chain for new vinyl records and helps them coordinate the comings and goings of parts and supplies. The adjective analogue may help capture the novelty of vinyl records as a medium, but it cannot fully account for the work of making them, especially when considering the various processes together and the cooperation between them. Both the manufacture and trafficking of records today entail what I elaborate in the following section as post-digital media logistics. The production of vinyl records entails the movement of material parts and supplies that cannot be digitized, but the coordination required between workers at the various stages of record production is no longer possible without digitized communication such as tracking numbers for deliveries and the use of the internet for file transfers as well as email and messaging.

At the same time, the three primary stages of vinyl production, when taken in isolation, each reveal limits of digital technology. Much of the communication between workers along the supply chain for new records has gone digital, but the actual work of making records has not and in most cases cannot. Mastering for vinyl requires a lathe equipped with a cutting stylus made from ruby or sapphire, and when mastered discs arrive at an electroplating facility, there is nothing digital in the silver sprayers and nickel baths used to render them into a form durable enough for fabrication. Finally, once the metallic plates are shipped to a pressing plant, synthetic plastic pellets are waiting to be melted and pressed between them with thousands of pounds of symmetrical, hydraulic pressure. Digital technology has been utilized to improve some of the monitoring and troubleshooting of a record press, but the core operations at each primary stage are essentially immune to digitization.

The tools and supplies used to make records can be squarely categorized as digital or not, and most of the vital ones are not. The labour of making records, meanwhile, has evolved into a historically revealing hybrid of craft work and industrial manufacturing. It is not that the finished product has changed – in fact just the opposite – but impressions of the product have changed. Digital technology and craft work are not mutually exclusive, but the work of records' fabrication has acquired new appreciation as non-digital, akin to the pleasures of listening to vinyl. The work of mastering, plating, and pressing records has come in for reevaluation, and in many ways, contemporary perspectives toward the labour of making records mirror attitudes about the product itself. Vinyl records have always been mass produced, and even boutique plants today need to press hundreds of copies to break even, yet every new record pressed now embodies the type of authenticity traditionally reserved for the fruits of handmade goods. Since vinyl replaced shellac as the default medium for popular music commodities, during World War II, the labour processes of making records have remained remarkably consistent; however, the meanings we ascribe to jobs are no less contingent, contextual, and historically specific as the meanings we ascribe to media. 'Few mass produced consumer goods signify authenticity to today's purchasers so richly as a vinyl record', and it follows that the labour of making records garners new types of respect, appreciation, and admiration (Barry, 2014). When it comes to vinyl manufacturing today, the new bleeding between traditionally opposed categories involves craft work and industrial labour, not analogue and digital technology.

A record press in motion may look nearly identical to one 50 years ago, but the people who have started up new pressing plants have infused the same old process 'with a new set of meanings and values that underlie the work they do' (Ocejo, 2017: xviii). Even when the press is running, the vibe at a record plant feels more like a workshop than a factory (see Figure 2.2). I have visited Smashed Plastic Record Pressing in Chicago at least half a dozen times, and more often than not, the owners have welcomed someone (else) wanting to come and see what they do and how it's done – sometimes it's a musician or someone from a label or from a company at another point along the supply chain, other times it's been reporters or friends or curious neighbours. Record plants are worth visiting in ways they were not 50 years ago. The process of making records may be essentially the same as it was 50 years ago, but it is more interesting now than it was then.

The number of record pressing plants in the US doubled during the 2010s, with a dozen opening from Oregon to South Florida.[6] The flowering of smaller, independently owned record pressing plants 'are part of a groundswell of small-batch "artisanal"' light manufacturing businesses that have recently emerged in the United States' (Ocejo, 2017: 54). Artisanal manufacturing sounds like an oxymoron, one I try to capture by describing contemporary record production as 'mass craft'. In historical terms, there is an irony in the development of an artisanal approach to manufacturing, in that the Arts and Crafts movement formed in Britain as a response to the same industrialized mode of production that is now, on the other side of digitization, being redeemed in worksites like record pressing plants.[7] Industrial production is distinguished from craft by the scale of operation and a reliance on machines rather than handiwork, yet the new spate of record pressing plants in the US tend to embody artisanal values. Much like the craft distilleries Richard E. Ocejo has discussed, record plants 'create and promote a sense of authenticity, or the idea of a product as full of integrity, truth, and real-ness [and they] promote a strong sense of localness in terms of where they source their ingredients [and] how they use place as a basis of their brand's identity' (Ocejo, 2017: 54). Craft brewers may be the iconic example of artisanal manufacturers in the US, and it is not a coincidence that the refrigerator in Smashed Plastic's listening room is stocked with free samples from Chicago-based microbrews.

From its logo on down, Smashed Plastic's hometown is central to its ethos and its business, and while two of its key 'ingredients' cannot be sourced locally, the vinyl material and stampers, the music can be, and the reasons that Smashed Plastic targeted local bands and labels exclusively at first were ethical as well as economical. The labour processes of making records may be the same as they were 50 years ago, but one way the pressing sector of vinyl production has changed is its new orientation toward the local. Plenty

Figure 2.2: The listening room at Smashed Plastic Record Pressing feels more like a lounge than a factory

Source: John Lombardo, Smashed Plastic co-founder/owner

of music is still pressed abroad, most (in)famously at GZ Media near Prague, chided by some competitors as 'the Walmart of vinyl pressing plants', but after the wave of new plant openings, independent labels can be more selective about where they send their records to be pressed.[8] For instance, former Merge Records production manager Katy Clove told me that she tried to send music to a press located in the city or region where the artists reside or recorded, and it is not uncommon for a label to now list the pressing plant among the credits on a new album's Bandcamp or Discogs page, particularly when the plant's location is consistent with the music's origins or aesthetics.[9]

Vinyl manufacturing today occupies 'a middle ground between artisanal and industrial modes of production' (Dudley, 2014: 128). Historically, the primary difference between the two modes is that they produce goods that are made by hand or machine. In our post-digital world, however, 'the cultural distinction between "handmade" and "machine made" has become increasingly blurred' (Dudley, 2014: 118). In some ways, vinyl's analogue credibility as a medium overshadows its mechanical production. Susan Luckman notes that, on the other side of digitization, 'seventy years since the Frankfurt School railed against "the culture industry's" commodification and standardization of all art, the analogue becomes Othered, different, desirable. This is particularly true of the handmade which is marked by its solid oneness in the world' (Luckman, 2013: 251). Amid a sea of readily accessed digital content, vinyl records may be 'Othered, different, desirable', but they are not handmade. In a canonical account of craft (written during vinyl's heyday), David Pye distinguishes between 'workmanship of risk' and 'workmanship of certainty', and pressing records more closely resembles the former. Work that qualifies as 'craftsmanship' employs 'any kind of technique or apparatus in which the quality of the result is not predetermined, but depends on the judgment, dexterity and care which the maker exercises' (Pye, 1968: 20–22). There is nothing certain about the quality of the smashed plastic that emerges from a record press. A machine may technically press the records, but there's no guarantee they'll all pass muster. In his ethnography of 'old jobs in the new urban economy', Ocejo notes that craft distillers of spirits insist that 'distilling is not that hard, if you follow the basic process' (Oceja, 2017: 301); however, they also insist that 'whether it's gonna be good is something else' (Oceja, 2017: 61). The studio manager at an esteemed mastering studio in Chicago captured the uncertainty that has always animated vinyl manufacturing, when he told me: 'any record pressed well is a minor miracle' (Nicolaou, 2019). Along the supply chain for new vinyl records, record pressing plants are the workplaces where the hybridity of artisanal manufacturing is most pronounced. Record pressing may still technically qualify as mass production, but it is now shot through with the authenticity and uncertainty of craft work.

Another uncertainty of vinyl manufacturing involves the elaborate and time-sensitive coordination of the supply chain. A well pressed record qualifies as a 'minor miracle' because of how much can go wrong before stampers reach a plant as well as once they do. When I visited a press for the first time, it surprised me that the workday at Brooklyn Phono started with a staff meeting, attended by every employee and lasting more than an hour before anyone fired up the presses. After my first couple of visits to Smashed Plastic, I learned to expect the workday to start in the listening room, where the owner-proprietors would review orders and plan the day's or week's schedule with the office manager (more often working remotely than in-person, even before the pandemic) and shift workers. Cut lacquers and stampers are 'touched multiple times', during their creation and then transport, and 'there's a multitude of things that can go wrong at any step along the way in making a vinyl record' (Hull, 2018). Befitting the specialized equipment and rarified work at each point of the process, mistakes made at one station can be difficult to fix at a subsequent one. And most confounding of all, sometimes 'nobody did anything wrong, it just didn't turn out great' (Hull, 2018).

Unlike digital formats, every record is imperfect. Each record pressed has always been technically less detailed than the one pressed before it, demonstrating 'that there indeed is such thing as a "unique copy"' (Bartmanski and Woodward, 2015: 33). By an ironic turn of events, the late modern revival and iconic consecration of mechanically reproduced analogue records neatly contradicted Benjamin's vision about mechanical reproduction sounding the death knell of art's aura (Bartmanski and Woodward, 2015: 32). The format fetish of vinyl's newly perceived 'analogue aura' permeates the supply chain; but at these work stations, that fetish involves the specific complexities and challenges of 'making a great record', not the pleasures of hearing one.[10] As one *Making Vinyl* panellist insisted, the digital dissatisfaction or 'disconnect being salved by vinyl' doesn't only flow from 'the experience of the product, but from the labour behind it' (Zapor-Gray, 2018). Marxian critics of capitalism may emphasize the work embodied in the consumption of any commodity, but in the case of vinyl, the balm of an analogue idyll can sooth the workers' digital wounds, too.

The post-digital logistics of vinyl's contemporary supply chains

When I attended the second annual *Making Vinyl*, held in Detroit during October, 2018, over the course of the conference's two days, it became clear that shipping was a stressor widely shared among the attendees. During a panel of new plant proprietors (seven white men in their thirties and forties, three with beards), one speaker assured the audience that 'no one is getting rich from this', and then another quipped, to knowing laughter, 'except FedEx and UPS'. Few facilities are equipped to handle more than one of vinyl production's three intricate operations, so delicate intermediate materials are usually transported from station to station. Profit margins are razor-thin for every phase of vinyl manufacturing, which only raises the stakes of efficiently delivering parts and materials between them, and any production delays can be especially costly for goods with release dates scheduled months in advance. During a *Making Vinyl* session devoted to 'Standards and Quality', a call for standardized packaging protocols struck a chord. Discussion zeroed in on transport between shops, and horror stories were shared about heavy boxes stacked too high and missed deliveries of mastered lacquers left to melt in the summer sun. Someone relayed the lore of a bicycle company having reduced damage 80 per cent by mislabelling their lean rectangular boxes as flat-screen TVs, and everyone acknowledged the inadequacy of 'FRAGILE: RECORDS' as a label to alert the uninitiated. Near the end of the session, a suggestion to invite representatives from the US Postal Service and leading shipping companies to the next *Making Vinyl* met with nods, murmurs of affirmation, and applause (a couple did attend the following year).

Once a vinyl record has been made, the audio information etched in its grooves cannot be streamed or downloaded, so any distribution and sales entail further physical relocation. Today the majority of vinyl is purchased online, virtually all of which is then mailed from sellers to buyers (a substantial and growing portion of records bought online are shipped internationally). Listening to vinyl requires far less energy than streaming, but internet sales have delocalized traffic and dramatically increased the resources devoted to shipping records. Like so many 'alternative practices in the age of the Anthropocene', vinyl's supply chains retain 'a dependency on mainstream architectures from global logistics to data centers and the perpetuation of an international division of labor' (Lovink and Rossiter, 2018: 7). Making and selling records today are both reliant on shipping, and it turns out that the revived vinyl economy is every bit as conditioned by logistical infrastructures as those digital formats to which records purportedly offer an alternative.

Vinyl is the leading case study in David Sax's popular book, *The Revenge of Analog* (2016), and Sax's formulation that 'digital [technology] helped save the very analog record it nearly killed', can be applied to the centrality of the internet for contemporary vinyl commerce as well as to the format's rearticulated appeal (Sax, 2016: 11). Ironically, vinyl's new value is enhanced by its perceived distance from the same technology appreciated, by buyers and sellers alike, as fuelling the contemporary vinyl marketplace. While vinyl's commercial infrastructure has migrated online, the format's materiality tethers it logistically to non-digital media. Chief among vinyl's logistical media are mail services. Every record made entails the transport of materials and supplies, and every online sale of a record occasions its packing and delivery.[11] Secondhand sales aren't tracked by the industry, and despite the decade of booming sales of new vinyl, 'the bulk of records traded, sold, and played today' are on at least their second owner (Sax, 2016: 20). This fact makes vinyl uniquely green among popular media, rivalled only by books, and the thriving online markets in used records underscore the centrality of shipping in the revitalized vinyl economy.

Amazon, of course, sells more new records than any other retailer, as well as a healthy share of used records; but all told, over two-thirds of new records are being sold by independent merchants, as well as most used records. Small shops (and individuals who flip records online) utilize the same, or at least the same sorts of logistical media – digital and non-digital – to facilitate their sales and organize their inventory as a behemoth like Amazon does at an exponentially vaster scale. Independent shops and individuals selling used records populate sites like eBay and Craigslist not to mention Discogs, the world's 'biggest and most comprehensive music database and marketplace' (Discogs, undated). Started as a 'hobby project' in 2000, the site is now 'like Wikipedia for music', with more than 600,000 users contributing to entries for over eight million artists and over 15 million recordings, but unlike Wikipedia it also lists nearly 75 million items for sale, over 50 million of which are vinyl records (Discogs, undated). Discogs has matured into an authoritative database as well as sales platform, and its roles as each are inextricable from the other, as purchases on Discogs help determine the going rate for records in shops and on other websites. Discogs embodies the centralizing, if not outright monopolizing force of 'platform capitalism' albeit on a smaller scale than an all-encompassing retailer like Amazon.[12]

No company has been more central to the logistical remediation of the vinyl marketplace than Discogs. Not only are millions of used as well as new records bought and sold via Discogs annually, but many of its more than seven million site visits per month are price checks from sellers as well as buyers. The time-honoured tradition of 'crate digging' has long been a source of exchange- as well as use-value for record collectors, and Discogs has upended the practice by providing access to prices and availability for novices alongside connoisseurs.[13] Discogs also affects the production of new records, directly as well as indirectly. No label worth its salt can afford to ignore the site's trove of data, and recently some labels have begun to float potential reissued releases with a Discogs entry, gauging interest before deciding whether or not to start the presses.[14] Discogs has become a factor in virtually all vinyl sales today, not just on the site itself, but offline as well as online.

Most record stores list at least some inventory on Discogs, even if just their more rarified fare, but their presence is dwarfed on the site by free agents selling records to help earn a living, to fund their own vinyl habit, or both. Shoppers can sort copies of a record for sale by price, condition, age, and location; users who keep a 'want list' are alerted as soon as any copies are posted for sale; and buyers and sellers are expected to leave reviews for one another. Many individual sellers will follow up repeatedly with buyers, imploring them (cheerfully) to leave a positive review, often offering (or withholding) their own in exchange (Discogs members are also assessed with a 'buyer's rating', based on sellers' feedback). In order to ensure that reviews remain uniformly positive, individual sellers often decide they must absorb the added expenses of refunds and return shipments. In the contemporary vinyl marketplace, risks and costs have been redistributed in patterns resembling those in other sectors of the retail economy that have been logistically remediated by 'platformization' (Nieborg and Poell, 2018; see also Kneese and Palm, 2020).

Dusty Groove emerged during the late 1990s as one of the first online retailers of records (or of anything else), and now it also has one of the largest and most established record stores among scores in Chicago. One of Dusty Groove's co-founders developed proprietary software to coordinate inventory and sales, customized to account for records' condition among other features unique to the format. As business has grown, their software has allowed Dusty to sell exclusively through their own website and avoid the costs – and data sharing – of partnering with Amazon or even Discogs.[15] Dusty Groove now maintains a staff of more than 20, several of whom are assigned to aspects of their online traffic, and the co-founder who developed the software works fulltime on maintaining and updating it (during slow hours, the number of customers browsing the shop floor is often matched by employees pulling records to fulfill online orders). Most smaller vinyl merchants cannot afford the labour and other costs of developing and maintaining their own software, and so they are forced to 'partner' with Amazon, Discogs, and other platforms that drive internet traffic, in the process sharing their data as well as profits.

Online record sales provide a rich example of post-digital commerce as well as culture, and one whose materiality exceeds the goods for sale. The online vinyl marketplace is only sustainable through the use of logistical media such as shipping boxes built specially for records alongside the tracking numbers used by buyers to remotely monitor those boxes in transit (the tracking number is post-digital logistical media *par excellence*). The labour of selling records has been logistically remediated, and the internet is a vital means of promotion and communication even for shops that don't sell online. Online sales can subsidize brick-and-mortar shops, and it has become a successful strategy for new record stores to establish their business online and then open a shop once sales reach a consistent level that can sustain the added costs. In addition to visibility and the prestige that comes with a desirable address, for merchants of bulky and breakable goods like records, a shop also functions as valuable storage capacity, even if the majority of sales continue to occur online.

The utilization of digital technology by distributors, merchants, and especially online storefronts has delocalized vinyl sales and dramatically increased the labour and resources expended for trafficking. Vinyl is noteworthy as an analogue alternative amid digital ubiquity, but digital media are vital to sustain the supply chains for the production of new records as well as the markets for used ones. Vinyl enjoyed a few decades as the default medium for popular music, when records were accepted by consumers rather than actively embraced, but now vinyl's revival gives the lie to Benjamin's famous critique of art and its mechanical reproduction: records are mass produced and industrially manufactured, yet they are now prized for their authenticity, for what Benjamin would call their aura. This rearticulated appeal of vinyl records, as an analogue exception in a digital world, makes them an ideal case study for demonstrating the centrality of post-digital logistical media to the production and distribution of popular culture in most any format today, whether its experienced offline or online.

Analogue labour idyll?

Documenting the work processes for making and selling vinyl records can further our understanding of the complex contemporary relationships between labour and digital technology. It remains impossible to digitize the primary processes of record manufacturing, yet coordinating the supply chains for new records relies on digital logistical media, such as the tracking number for deliveries of essential non-digital parts like masters, stampers, and the vinyl material itself. Meanwhile, vinyl's rearticulated desirability as a commodity may be animated by its analogue exceptionality amid a sea of digital content, but the vinyl marketplace is thoroughly digitized, reliant on sales platforms, search engines, and social media. If the new aesthetic experience of listening to records qualifies as post-digital, then no less does the contemporary work of vinyl's production and distribution.

The brief career of 'digital labour' as a critical term underscores the continued utility of documenting the work of making and moving non-digital media like records. During the late 2000s and early 2010s, the term 'digital labour' swept into media studies.[16] The new compound noun emerged as a critical term meant to expose the 'leisure-driven, unpaid activities of social media users as unremunerated forms of work contributing to Internet companies' profits' (Gandini, 2021: 369). Since then, the capitalization of user-generated content has metastasized, and the term 'digital labour' has lost its edge, expanding into 'a sort of umbrella term that is increasingly delinked from its origins as a critical Marxist stance on labour and value' (Gandini, 2021: 370). Now the term is being used to describe any endeavour, paid employment as well as appropriated leisure, that includes the use of digital technology. The problem is that 'it is now almost impossible to untangle the "digital" prefix from the "labour" activity', and the 'blanket use of the expression "digital labour"' has left it 'unable to make a clear distinction between Facebook, Twitter and YouTube, on the one hand, and the likes of Uber and Deliveroo on the other' (Gandini, 2021: 376–377).[17] Alessandro Gandini (2021) argues that 'digital labor' has devolved into 'an empty signifier', noting that, because 'the two terms – digital *and* labor – are increasingly inseparable', there is a corresponding 'increasing impossibility of debating labor and digital technology as two separate components [of] critical academic scholarship' (Gandini, 2021: 370). Relative to the much ballyhooed pleasures of records' rearticulated consumption, the labour of vinyl production and distribution remain understudied, and one reason vinyl makes for an illuminating case study of contemporary labour is because it simultaneously highlights some of the limitations as well as the now-indispensable capacities of digital technology when it comes to the creation and circulation of analogue cultural commodities.[18]

Gandini (2021: 377) observes that 'the presence of a "digital" component in labour today is nothing special', although maybe for vinyl it is. When I lighted on record makers and merchants as research subjects, I was tempted to adopt the phrase 'analogue labour' as a way to distinguish them from workers along the supply chains for other forms of popular culture, namely digital ones. Contemporary vinyl manufacturing and trafficking are each worthy of their own book, but there are advantages to analysing them together. For starters, it allows me to follow recorded music from the mastering studio through to the point of sale and, in the case of used records, resale. By adopting a both/and purview for vinyl labour, as mentioned already, a digital disconnection of sorts emerges between the contemporary manufacture of records and their trafficking. The phrase 'analogue labour' could arguably apply to making records, but not to selling them, and rather than trying to find ways to unify these two modes of vinyl labour, I find it more fruitful (and faithful) to tease out the differences alongside the commonalities.

One upshot of labelling contemporary vinyl production as 'analogue labour' is that it redirects attention away from the experience of consuming a celebrated analogue artifact and toward the experience(s) of making it. The admittedly unnecessary neologism (enough new labours already!) can also remind us that the revived markets for new records are preserving forms of pre-digital labour as well as leisure, modes of production and distribution as well as consumption now identifiable as post-digital. In an historical irony, the type of work being preserved and lauded for its rewarding artisanal challenges and sensibilities is essentially industrial manufacturing, which before digitization was seen as the very mode of degrading work responsible for superseding its more rewarding predecessor, namely craft work. In the case of vinyl records and their revived economy, it is now manufacturing being lauded for providing opportunities for labour that are more enriching, somehow more authentic than any of myriad types of work reflected back to us in the glow of a screen.

Notes

1 https://www.discogs.com/company. I estimate the volume of used records by subtracting the number of records for sale that are described as being in 'mint' or 'near-mint' condition, and therefore presumed new, from the total number of records for sale. The figure is conservative, since many used records are listed in 'near-mint' condition.

2 For represses and reissues, it's often years or decades.

3 On 6 February 2020, the Banning, California, factory of Apollo Master Audiodiscs was destroyed in a fire, leaving one supplier of blank lacquers globally. A month later, the fire's impact on new vinyl production was overshadowed – and delayed – by pandemic lockdowns (see Yoo, 2020).

4 A single company, Thai Plastics and Chemicals [TPC], was credited as recently as 2015 with supplying upwards of ninety percent of the vinyl used to make records (see Fact Staff, 2015).

5 On the emergence of digital labour, see Burston, Dyer-Whiteford, and Hearn (2010) and Scholz (2013).

6 See Total Sonic Media (undated) for a list of pressing plants, last updated 4 April 2024.

7 On the Arts and Craft movement, see Luckman (2013).

8 Dean Reid, who launched Canada Boy Vinyl, quoted in Bitti (2016).

9 Clove (2018). And to offer just one example, the Bandcamp description of Daniel Bachman's 2021 album *Axacan* begins: '"Axacan" is the fourth album from Daniel Bachman for Three Lobed and his first album in three years. It was recorded in 2020 at various locations in Virginia. It will be pressed on two 140 gram 12" records in Virginia by Furnace [Record Pressing, located in Alexandria]'. https://threelobed.bandcamp.com/album/axacan (It's also worth noting that, since then, Furnace has been purchased by Metallica (see Smith, 2023c).)

10 This phrase was a refrain at the second and third annual *Making Vinyl* conferences I attended, in Detroit, 2018, and Los Angeles, October, 2019. And in this context the 'greatness' of a record refers to its sound quality, regardless of one's opinion of the music on it.

11 The exceptions are records sold online to local buyers who pick them up from sellers.

12 See, among many others, Srnicek (2016: 92ff).

13 'You ask me why I carry cash around with me all the time. / Don't you see all these yard sale signs?/ When I flip this first edition for a cool grand, / Then maybe you'll understand ...' Wooden Wand, 'Mexican Coke', *Clipper Ship* (Three Lobed Recordings, 2017). Discogs has made these kinds of finds more scarce, even at flea markets and yard sales, where sellers can look up the going rate with a quick search.

14 Another sign that a reissue is coming is when prices on Discogs drop for rare original pressings.

15 For a compelling account of Dusty Groove focused on their offline procurement of inventory, see Danielle Beverly's 2019 award-winning documentary, *Dusty Groove: The Sound of Transition*.

16 On the emergence of 'digital labour' within media studies, see Burston, Dyer-Whiteford, and Hearn (2010) and the special double issue they edited; and Scholz (2013). See also Palm (2011).

17 Gandini notes that there are 'exceptions, such as Srnicek, 2016'.

18 As far as I know, there remains no sustained study – popular or scholarly, historical or contemporary, ethnographic or otherwise – of the labour processes of vinyl production.

Record stores, conversely, are a popular *mis en scene* for novels and movies, most famously *High Fidelity*, as well as the subject of books and articles. See, most recently, Arnold et al (2023).

References

Arnold, G., Dougan, J., Feldman-Barrett, C., and Worley, M. (eds) (2023) *The Life, Death, and Afterlife of the Record Store: A Global History*. Bloomsbury Academic.

Barry, E. (2014) 'Digilog Culture and the Vinyl Revival of the Early 21st Century', *American History Now: Explorations in Digital Curation*, [online] 30 April, Available from: https://americanhistorynow.org/index.html%3Fp=475.html [Accessed 20 April 2024].

Bartmanski, D. and Woodward, I. (2015) *Vinyl: The Analogue Record in the Digital Age*. Bloomsbury Academic.

Berry, B. and Dieter, M. (eds) (2015) *Post-digital Aesthetics: Art, Computation, and Design*. Palgrave-Macmillan.

Beverly, D. (2019) *Dusty Groove: The Sound of Transition*. Petunia Productions.

Bitti, M.T. (2016) 'Can a Vinyl Comeback Sustain Two Players in the Canadian Market?' *Financial Post*, [online] 14 July, Available from: https://financialpost.com/entrepreneur/growth-strategies/can-a-vinyl-comeback-sustain-two-players-in-the-canadian-market [Accessed 20 April 2024].

Blocker, D. (2018) 'Panel Discussion: New Vinyl Plants Fire Up the Presses', *Making Vinyl* conference, 1 October, Detroit, MI.

Burston, J., Dyer-Whiteford, N., and Hearn, A. (2010) 'Digital Labour: Workers, Authors, Citizens', *Ephemera: Theory & Politics in Organization*, 10(3): 214–221.

Clove, K. (2018) Interviewed by the author. 28 December, Durham, NC.

Discogs (undated) 'About Discogs', [online], Available from: https://www.discogs.com/about [Accessed 20 April 2024].

Dudley, K.M. (2014) *Guitar Makers: The Endurance of Artisanal Values in North America*. University of Chicago Press.

Eastes, S. (undated) 'The Persistence of Vinyl', *The Bitter Southerner*, [online] Available from: https://bittersoutherner.com/united-record-pressing#.XGeK2s9KhE6 [Accessed 20 April 2024].

Fact Staff (2015) 'US Vinyl Suppliers Struggling to Keep with Demand from Pressing Plants,' *Fact*, [online] 14 December, Available from: https://www.factmag.com/2014/12/15/us-vinyl-suppliers-struggling-to-keep-up-with-demand-from-pressing-plants/ [Accessed 20 April 2024].

Gandini, A. (2021) 'Digital Labour: An Empty Signifier?', *Media, Culture & Society* 43(2): 369–380.

Herrmann, T. (2015) 'Pressed to the Edge: Why Vinyl Hype Is Destroying the Record', *Fact*, [online] 7 May, Available from: https://www.factmag.com/2015/05/07/pressed-to-the-edge-vinyl/ [Accessed 20 April 2024].

Hull, S. (2018) 'Panel Discussion: Vinyl Mastering & Cutting Intricacies', *Making Vinyl* conference, 1 October, Detroit, MI.

Jaffee, L. (ed) (2017) Making Vinyl Detroit 2017 Program Guide, [online] https://makingvinyl.com/wp-content/uploads/2022/07/MakingVinyl_2017_brochure_web_sp.pdf [Accessed 12 April 2024].

King, A. (2022) 'Vinyl Outsells PlayStation in the UK – Next Challenge? Nintendo Switch', *Digital Music News*, [online] 7 September, Available from: https://www.digitalmusicnews.com/2022/09/07/vinyl-outsells-playstation-in-the-uk/ [Accessed 12 April 2024].

Kneese, T. and Palm, M. (2020) 'Brick-and-Platform: Listing Labor in the Digital Vintage Economy', *Social Media + Society*, July–September: 1–11.

Luckman, S. (2013) 'The Aura of the Analogue in a Digital Age: Women's Crafts, Creative Markets, and Home-Based Labour After Etsy', *Cultural Studies Review*, 19(1): 249–270.

Lovink, G. and Rossiter, N. (2018) *Organization after Social Media*. Minor Compositions.

Mazierska, E., Gillon, L., and Rigg, T. (eds) (2019) *Popular Music in the Post-Digital Age: Politics, Economy, Culture and Technology*. Bloomsbury Academic.

Nicolaou, A. (2019) Interviewed by the author. 1 March, Chicago.

Nieborg, D. and Poell, T. (2018) 'The Platformization of Cultural Production: Theorizing the Contingent Cultural Commodity', *New Media & Society*, 20(11): 4275–4292.

Ocejo, R. (2017) *Masters of Craft: Old Jobs in the New Urban Economy*. Princeton University Press.

Palm, M. (2011) 'Labor's New *Empire*', *Journal of Communication Inquiry*, 35(4): 433–438.

Pye, D. (1968) *The Nature and Art of Workmanship*. Cambridge University Press.

Rutkowski, S. (2018) Interviewed by the author, 1 October, Detroit, MI.

Sax, D. (2016) *The Revenge of Analog: Real Things and Why They Matter*. Public Affairs.

Scholz, T. (ed) (2013) *Digital Labor: The Internet as Playground and Factory*. Routledge.

Smith, D. (2023a) 'UK Music Industry Sales Cracked $2.4 Billion During 2022 As Vinyl Outsold CDs For the First Time Since 1987, Report Shows', *Digital Music News*, [online] 10 January, Available from: https://www.digitalmusicnews.com/2023/01/10/uk-music-industry-revenue-2022/ [Accessed 12 April 2024].

Smith, D. (2023b) 'U.S. Recorded Music Revenue Approached $16 Billion In 2022 Amid Continued Vinyl and Streaming Growth, Report Shows', *Digital Music News*, [online] 9 March, Available from: https://www.digitalmusicnews.com/2023/03/09/music-industry-revenue-2022-us/ [Accessed 12 April 2024].

Smith, D. (2023c) 'Metallica Purchases Entire 70,000-Square-Foot Record Pressing Plant Amid Vinyl Resurgence', *Digital Music News*, [online] 14 March, Available from: https://www.digitalmusicnews.com/2023/03/14/metallica-furnace-buyout/ [Accessed 21 April 2024].

Srnicek, N. (2016) *Platform Capitalism*. Polity Press.

Total Sonic Media (undated) 'Vinyl Record Pressing Plants,' *Total Sonic*, [online] Available from: https://www.totalsonic.net/vinylplants.htm [Accessed 20 April 2024].

Yoo, N. (2020) ' "Devastating" Manufacturing Plant Fire Threatens Worldwide Vinyl Record Supply', *Pitchfork*, [online] 7 February, Available from: https://pitchfork.com/news/devastating -manufacturing-plant- re-threatens-worldwide-vinyl-record-supply/ [Accessed 14 December 2020].

Zapor-Gray, J. (2018) Panel Discussion: The Power of Crate Diggers, *Making Vinyl* conference, 2 October, Detroit, MI.

3

'We Don't Have Wi-Fi, Talk to Each Other': Digital Disconnection in the Hospitality Industry

Alexandra Kviat

Introduction

'WE DON'T HAVE WI-FI, TALK TO EACH OTHER'. This sign, first spotted in European and North American cafes and bars about 12 years ago, has become a long-running internet meme and an inspiration to many entrepreneurs across the globe – so much so that it is now sold ready-made on Amazon, eBay, and Etsy. It is no wonder that one of the most prominent symbols of disconnection culture emerged within the hospitality industry. Cafes, restaurants, pubs, bars, and other businesses representing the popular concept of the 'third place', introduced by sociologist Ray Oldenburg (1989) as the opposite of the home (the 'first place') and the office (the 'second place'), are the hubs of economic, social, and cultural activity that form an integral part of everyday life. By bringing people together in a relaxed, informal, and cosy setting, third places provide the social glue for urban and rural communities, foster civic engagement, and strengthen the personal and communal sense of identity and belonging (Oldenburg, 1989; Putnam, 2000; Jeffres et al, 2009). At the same time, the majority of such establishments are commercial enterprises whose day-to-day operations are focused on the production and consumption of food and drinks and guided by the imperative of economic sustainability. Over the past two decades, the advancement of wireless and mobile technologies, accompanied by the growth of non-office-based working, has dramatically changed the functions and design of third places (Hampton and Gupta, 2008; Henriksen and Tjora, 2018). Traditionally used for refreshment and socializing, they have been yielding to the growing pressure of today's 'digital by default' culture (Kuntsman and Miyake, 2019), offering their customers free Wi-Fi access, installing additional power outlets, and rearranging seating to better accommodate the needs of laptop users (Schulte, 2011). Nevertheless, this new norm is being challenged by an increasing number of entrepreneurs who choose to limit or fully ban the use of digital technologies on their premises (Famurewa, 2016; Kadet, 2017; Herd, 2018; Stabiner, 2018).

Despite attracting a lot of public attention with their promise to 'bring back the art of conviviality' (Herd, 2018), such places, commonly described as 'tech-free', 'screen-free', 'laptop-free', 'smartphone-free', 'power-free', 'no Wi-Fi', 'no-fi', 'lo-fi', 'low-tech', or even 'anti-internet', have not yet been the subject of scholarly discussion. Does cutting off Wi-Fi really foster offline conversation, and why is it important for hospitality entrepreneurs? What else do they do to prevent customers from using digital technologies, and what motivates these efforts? How do disconnected third places affect wider society? This chapter, informed by an interdisciplinary body of literature including media studies, sociology, cultural history, and consumer and service research, aims to address these questions by exploring the range and meaning of digital disconnection policies implemented in the third places of the Global North and by challenging some popular assumptions about their possible effects. The empirical data comes from a desk study conducted in 2020 in two strands: 1) a survey of disconnection policies used by a sample of 60 hospitality businesses operating in the United Kingdom, the United States, Canada, Australia, Denmark, and France, and 2) a discourse analysis of media publications discussing the topic of digital disconnection in third places. The sources included news articles, feature stories, op-eds, blog posts, company website and social media content, and customer reviews on Google Maps and TripAdvisor.

Like other contributions in this collection, this chapter offers a critical analysis of the analogue/digital dialectic in contemporary culture, while also shedding light on how the discourse on technology relates to other issues and conflicts associated with hospitality entrepreneurship and public space. I begin by situating disconnected third places in their historical and theoretical context and then provide an overview of the identified disconnection policies, before discussing in more detail their key impacts on business and society.

The history of disconnected third places

In June 2005, a small Seattle-based independent cafe called Victrola Coffee and Art found itself on the pages of *The New York Times*. To the slight disappointment of Victrola's owners, it was not their locally renowned coffee that sparked the national press's interest – instead, the article focused on their decision to shut down their free Wi-Fi, even if only on weekends. With over 3,500 cafes offering a Wi-Fi service to US customers at that time (and 5,000 more providing it for a fee), it certainly was a move against the grain:

> Victrola started providing free wireless access two years ago after customers asked for it. As in hundreds of other cafes, the owners hoped it would encourage regulars and infrequent patrons to buy more food and drinks. But there was also a disadvantage, staff members said: the cafe filled with laptop users each weekend, often one to a table meant for four. Some would sit for six to eight hours purchasing a single drink, or nothing at all. Even worse, when lingerers were confronted, they were bellicose. … But Ms. Strongin and her staff said they were more concerned that the cafe had turned into a place devoid of sound except the light clack of keys, not the focus of lively interaction that she and her husband had intended. So last month the cafe discontinued the free service on Saturdays and Sundays – and so far it has proved to be a sound business move. Weekend revenue is up and more seats are filled. (Fleishman, 2005)

As the first known example of a disconnected third place put under the media spotlight, Victrola's case gives an insight into the background that eventually led more entrepreneurs to question the 'digital by default' norm, spearheaded in the early 2000s by large corporate chains like Starbucks and Panera Bread. Here we see how Wi-Fi access, initially envisioned by cafes as a complimentary service meant to help them attract and retain customers, quickly became normalized and taken for granted. Not only did Victrola's owners, like many other proprietors, feel that their investment in digital technology did not pay off economically, but they also found it counterproductive to the social environment they were trying to create. Once the free Wi-Fi policy was introduced, it reportedly took only a year to drain all life from what used to be the neighbourhood's social hub. Hence, the subsequent U-turn was interpreted as Victrola's attempt to 'take back its culture' from unsociable and inconsiderate customers who took advantage of the free Wi-Fi. The 'laptop squatter' stigma (Guynn, 2010) was also frequently mentioned in other news reports and opinion pieces of the time (Schulte, 2011).

On closer examination, the story of Victrola represents two fundamental tensions that continue to galvanize the public debate on digital disconnection in the hospitality sector to this day. The first tension, between unmediated and mediated communication, stems from the widespread perception of information and communication technology, or ICT, as an ever-increasing threat to social cohesion and particularly to the third place, conceptualized as a realm where such cohesion is produced through face-to-face conversation (Oldenburg, 1989: 49). Long before the appearance of mobile devices and wireless networks, sociologists and media scholars were keen to hold telecommunications accountable for the alleged loss of community and the rise of privatism and social atomization (Meier, 1962; Sennett, 1977). The expansion of personal computers and home internet access from the late 1980s to the early 2000s further widened the perceived gap between traditional, physically tangible forms and spaces of socialization and leisure, typically described as a dying breed in need of preservation (Oldenburg, 1989; Putnam, 2000; Gumpert and Drucker, 2005), and 'the hypermaterialistic synthetic connections of computer networks where the body, technology, and community are reduced to the hallucinatory metaphors of cyberspace' (Boyer, 1992: 126). This clear-cut opposition between the real and the virtual, very common for academic and public discourse before Web 2.0, can be illustrated with a *Washington Post* review of a local bar, titled 'The Saloon: A True Chat Room': 'Communication among people is one of the most important things', says the owner of the Saloon … . There's no television to turn patrons into staring, slack-jawed zombies. … Instead, he relies on one of the oldest forms of entertainment: human interaction (Hahn, 2004).

Even though this review does not directly discuss digital disconnection (it was published a year before Victrola made the news), it is remarkable how it reclaims the online chat room concept, implying the ersatz nature of virtual third places, and condemns television using the same rhetoric that was later applied to laptops, tablets, and smartphones. The further we delve into the origins of this discourse, the more evident it becomes that the presence of ICT in third places – especially in cafes, traditionally used as reading and writing spaces – has never been unproblematic. Contrary to the popular idealized view of late seventeenth- and early eighteenth-century English coffeehouses as spaces of lively public discussions, their patrons were routinely criticized by their contemporaries for an 'inordinate appetite' for newspapers (Cowan, 2005: 235). The following quote from Nikolay Karamzin, a Russian writer who visited London in 1790, reads strikingly similar to the scene described by Victrola's owners over two centuries later: 'I have dropped into a number of coffeehouses only to find twenty or thirty men sitting around in deep silence, reading newspapers, and drinking port. You are lucky if, in the course of ten minutes, you hear three words' (Cowan, 2005: 86).

The second tension apparent in Victrola's story is the conflict between the social and economic aspects of hospitality entrepreneurship. As has been observed by communication scholar Stephanie R. Schulte (2011), the debate about the use of digital technologies in the public realm is essentially a debate about whether public spaces should be designed for economic productivity or civic engagement. This dialectic between market logic and social values is particularly tense in commercial third places. Privately owned yet publicly accessible, they occupy an ambiguous semi-public niche on the continuum between fully public and fully private spaces. There was reason for Victrola's owners to make a point about not being driven solely by profit but rather trying to protect the cultural ideal of the third place as a social hub where people meet and interact with one another. Although semi-public consumption environments have always been an inherent part of the public realm, they are often dismissed as mere manifestations of the ever-increasing privatization and commercialization of public space. However, there is vast empirical evidence that commercial third places perform an important social and cultural function in the lives of urban and rural residents (Jeffres, 2009; Hickman, 2013; Jones et al, 2015). Furthermore, several studies have indicated a closely intertwined, if not entirely inseparable relationship between the social and commercial aspects of hospitality (Lugosi, 2007) and, more specifically, between the social and cultural ambitions and economic motives of entrepreneurs operating in this sector (Hubbard, 2017). And yet, the discourse on disconnected third places is firmly structured around the reductionist opposition between profit maximization and social good just as much as it is framed by the value-laden distinction between mediated and unmediated communication.

Both tensions became more acute in the early and mid-2010s, leading more hospitality businesses across the Global North to opt out of the 'digital by default' social contract with their customers. First called a shift, then a trend, this development resulted, on the one hand, from the global recession of 2007–09. Job losses and casualization of labour, along with public service cutbacks and rising housing and rental prices, prompted more people to use third places as temporary offices (Hampton and Gupta, 2008; Henriksen and Tjora, 2018), thus challenging the already blurred line between post-industrial work and leisure (Lewis, 2003). While pre-recession media publications suggested that people who work from third places 'don't do so because they are too cheap to rent an office', but because they 'enjoy being around other people' (Anderson, 2006), this 2009 *New York Times* article painted a much darker picture:

> In New York City, where the unemployment rate remains at 10.3 percent, the jobless have started leaning hard on coffee shops and bookstores to get out of their tiny one-bedrooms and away from their annoying roommates. In these harsh, career-vanishing times, the members of this laptop brigade do everything they can to re-create the office environment they no longer have to complain about. (Dolan and Daniel, 2009)

The war metaphor used in this fragment ('laptop brigade') reflects the conflicting interests of this precarious clientele, commonly portrayed in the media as 'frugal customers' who 'consumed less and stayed longer' (Guynn, 2010), and hospitality business owners, who were themselves struggling to survive in a tough economic climate. By 2016–18, when the discourse on disconnected third places reached its peak, the hospitality sector had mostly recovered from the immediate effects of the 2007–09 credit crunch, but the trend towards remote, mobile and freelance working was still growing. At the same time, the shrinking of public infrastructure and the rising cost of living in the post-recession economy made commercial third places one of the very few available options for socializing and leisure outside home (Ferreira, 2017). While practice- and policy-oriented research was calling for the recognition of laptop workers as an economically valuable demographic (Di Marino and Lapintie, 2020; Mimoun and Gruen, 2021), the media continued to stigmatize them with new derogatory names (for example, 'hoggers', 'campers', 'rustlers', 'chancers', 'freeloaders', 'rats', 'hobos'), portraying them as pretentious parasites stealing revenue from business owners and space from other customers. Altogether, it increased pressure on third places to perform multiple, sometimes conflicting social functions while remaining profitable.

On the other hand, the disconnection trend was also boosted by the emergence and fast consumer adoption of new portable communication devices (first and foremost, smartphones), 3G and 4G networks and social media platforms. By allowing people to be simultaneously present in several contexts and situations, these new technologies blurred the once rigid distinction between material and digital space (Hampton and Gupta, 2008; Ridell and Zeller, 2013), thus giving rise to the culture of 'always on', 'anytime anywhere' connectivity. Since the middle of the past decade, disconnection as a personal and organizational response to 'the growing sense of being overwhelmed by the digital, in one's work or personal life', has been moving 'from the margins to the centre' (Kuntsman and Miyake, 2016). Disconnected third places, along with other spaces and practices of digital withdrawal (Kuntsman and Miyake, 2016; Stäheli and Stoltenberg, 2024), became part of a wider anti-digital discourse, largely fuelled by anecdotal evidence and the long tradition of moral panics about technology. From popular science books and think pieces (Turkle, 2011; Bowles, 2019) to countless internet memes, digital technologies are imagined as a hegemonic force encroaching into all spheres of life, disengaging people from each other and reality and creating a hollow simulacrum of sociality that bears no comparison to that of pre-digital times, including the relatively recent Web 1.0 past.[1] However, as was recently pointed out by media scholars Adi Kuntsman and Esperanza Miyake, disconnection should not necessarily be interpreted as a rebellion against digitization or, conversely, as a capitalist co-optation and repackaging of such protest into yet another consumable fad, for it is not a monolithic phenomenon but rather a continuum of practices, motivations and effects, 'many of which differ by cultural, social, linguistic and political contexts and are yet to be explored in depth' (Kuntsman and Miyake, 2019: 908). In that spirit, I now turn to a more detailed and contextualized discussion of disconnection policies implemented in third places and the various meanings attached to them by business owners and the media.

Disconnection policies: an overview

Disconnection in third places can be implemented through a diverse range of technology-focused measures and policies that control customer access to mobile and Wi-Fi networks, as well as the use of personal communication devices on the premises (Table 3.1). Network-focused policies do not explicitly forbid customers to use their communication devices – instead, they disrupt their connectivity to mobile and Wi-Fi networks. To block mobile signal, some proprietors line the walls and ceiling with conductive materials (such as silver foil, aluminium or copper mesh), which creates a Faraday cage effect. Another, more common approach is non-provision or limitation of access to Wi-Fi networks, which can be switched off permanently or during peak days and hours, protected by one-time passwords expiring after a certain period (usually between 30 and 120 minutes), or never installed in the first place. Contrastingly, device-focused policies openly restrict or prohibit customer use of personal communication devices through a range of partial and full bans. While restrictive policies limit the use of laptops, tablets and, less commonly, mobile phones to certain time intervals (for example, weekdays or morning hours), dedicated spaces (such as bar counters, communal tables, patio seats) or specific functions (such as camera mode only), prohibitive ones fully ban at least one and sometimes all three types of devices.

In addition, disconnected third places employ an array of socio-spatial instruments that serve as a softer alternative to direct control or as a supplement intended to amplify its impact (Table 3.2). Some of these indirect tactics aim to obstruct the use of digital technology through spatial

Table 3.1: Direct (technology-focused) instruments of disconnection

Control of networks	Control of devices	
• Blocking mobile signal • No Wi-Fi access • Time-limited Wi-Fi access	• Time-limited laptop and tablet use • Time-limited laptop use • Space-limited laptop use • Limited smartphone use (camera only)	• Laptop ban • Laptop and tablet ban • Mobile phone ban • All-device ban

Table 3.2: Indirect (socio-spatial) instruments of disconnection

Instruments	Obstructive	Constructive
Spatial	• No/few power outlets • Hidden Wi-Fi password • Uncomfortable/ limited seating • Loud music	• Open bar • Soft seating • Communal tables • Vintage décor • Theme art • Handwritten signs and notes • Analogue media • No TVs and gambling machines • Quiet/no music
Social	• (Non)verbal cues from staff	• Facilitation by staff • Social events • Conversation-starter cards • Board games • Self-service

design (no/few power outlets, no clear indication of Wi-Fi password, uncomfortable or limited seating, loud music) or social means (verbal and nonverbal cues from staff, for example, checking up on customers or cleaning their tables). Other tools take a more constructive orientation by fostering face-to-face interaction and immersing customers in analogue culture. Often used to enhance the impact of direct disconnection policies, such constructive tactics include spatial design solutions (from furniture and décor to soundscape) and social practices and interventions (for example, facilitation by staff, social events, conversation-starter cards, board games, and self-service).

The choice of instruments in each specific case is determined by the intended purpose of disconnection, which, in turn, depends on the type of business. For instance, disconnection policies implemented in cafes and restaurants mostly target laptops – relatively bulky objects seen predominantly as work tools and thus associated with 'squatting' and 'table-hogging', whereas pubs and bars tend to crack down on mobile phones – smaller and more ubiquitous devices, the epitome of the 'always on' culture. However, the same disconnection policy can change its meaning over time. Until the mid-2010s, earlier generations of mobile phones, used primarily for voice calling, were banned in some drinking establishments as a source of noise nuisance, the acoustic equivalent of smoking in public, though perhaps without quite the same negative health impacts (Kelly, 2009). With the arrival of smartphones and the resulting cultural and technological shift to the quieter but arguably more isolating practices of texting and scrolling, the mobile phone bans were rearticulated as a measure against a new threat to nightlife sociality – excessive silence rather than noise (BBC, 2016).

Furthermore, it is not uncommon for third place owners to rethink or reverse their disconnection policies in response to customer criticism (Hauser, 2018).

When asked about their motives, owners and managers of disconnected third places often start with a disclaimer that they are 'not Luddites or technophobes' (Barns, 2016). Indeed, they are by no means trying to uninvent the internet and gadgets, and some even emphasize that they actively use social media for business promotion. What they offer instead is typically described in softer, emancipatory terms such as 'refuge', 'relief', or 'break' (from digital technologies). Another common trope is de-emphasizing or denying economic goals while highlighting social and cultural ones, such as trying to 'increase social engagement' (Lewis, 2017) or 'revive the art of hospitality' (Herd, 2018), or even help customers 'take back control of their humanness' (Schwartz, 2019). Although it may be equally tempting to accept such claims at face value or dismiss them as a disguise for profit-seeking, it is important to be wary of the either-or trap when it comes to judging the motives behind hospitality entrepreneurship. The question therefore should be not whether digital disconnection in third places is intended to do social good or generate more revenue for their owners but rather what aspects of business (including but not limited to economic ones) it may affect and how; and what impact it may have on wider society.

Disconnection impacts on business and society

Whatever rhetoric is used by hospitality entrepreneurs to justify the use of disconnection policies, the range of their expected outcomes boils down to the following two effects. The first is an increase in sales, which can be achieved through speeding up table turnover and/or through qualitative changes in customer experience, brand positioning and promotion. The second is the fulfilment of the entrepreneur's non-economic ambitions, including self-realization and social and cultural change. However, profit-driven disconnection policies also affect wider society by entrenching certain stereotypes and assumptions. Before I discuss each of these effects in more detail, it is important to note that they are closely intertwined, that is, the same policy can have multiple impacts on a particular business.

Table turnover

From the 2005 story of Victrola to more recent cases, many hospitality entrepreneurs use disconnection to increase table turnover by discouraging customers from lingering and table-hogging. This issue primarily affects cafes – historically versatile spaces used, among other things, for relaxation and contemplation – and particularly small independent cafes, struggling to compete with corporate chains. The latter 'may not feel the pinch' because of their 'multibillion-dollar revenues and legions of grab-and-go customers' and hence can afford to be more digital-friendly (Stabiner, 2018). Not only do Starbucks, Costa, and other behemoths dominate the market, both spatially and financially, but they also set a norm where 'a $4 latte turns into a five-hour stay', which cannot be fulfilled by smaller proprietors (Niezgoda, 2019). This problem is particularly acute in cafes that serve meals, where table hoggers are believed to 'squeeze out the more lucrative lunch or dinner crowds' (Guynn, 2010). Common solutions to customer overstaying include removing or limiting access to Wi-Fi networks and partial or full bans of laptops and, less often, tablets. In contrast, smartphones and analogue media such as books, newspapers, and magazines are rarely seen as a threat to table turnover. While smartphones tend to be exempt from such bans because of their relatively limited usability, analogue media are typically perceived as a healthier, more natural alternative to laptops and tablets as they do not lure customers away from the physical here and now to the same extent, leading to hours of distracted scrolling. To quote the owner of Dough Lover cafe in Brighton, UK, 'When you open a computer, you move beyond a natural time cycle and lose touch with the environment around you' (Herd, 2018). This is where indirect obstructive tactics come into play:

> At Triniti [a Los Angeles cafe], Mr. Wynn offers free Wi-Fi, but after two hours a customer must have 'a face-to-face interaction' with an employee to get a new password. He relies on that interaction and the aromas from the kitchen to transform coffee drinkers into lunch customers, and has been gratified as the first wave of workers have started to order midday meals. (Stabiner, 2018)

It is not entirely clear whether disconnection always results in higher revenue, as this and many other anecdotal accounts contend, and what the actual proportion of inconsiderate customers among all patrons with laptops and tablets might be. According to ethnographic studies of people using third places for work purposes, this clientele does its best to abide by such unwritten etiquette rules as 'avoid large tables', 'make a purchase every two-three hours', 'share power outlets' (Henriksen and Tjora, 2018; Di Marino and Lapintie, 2020). Moreover, not everyone using their laptop or tablet in a cafe does so to take advantage of the owner's hospitality. For those visiting third places by themselves, personal communication devices also perform the role of security blankets keeping them company or protecting them from unwanted attention (Hampton and Gupta, 2008; Henriksen and Tjora, 2018; McCamley and Morland, 2021). Because solo customers are believed to be less profitable, disconnection often serves to turn them away on the pretext of 'ensuring space for families and groups of people' (Hauser, 2018), which can easily be perceived as a discriminatory policy (Dawson,

2019). Although the primary goal of anti-loitering measures is to protect the economic interests of business owners, they are also likely to perpetuate the 'laptop squatter' stigma and reinforce negative stereotypes about being alone in third places (McCamley and Morland, 2021).

Customer experience

For many third places – not only for cafes, as in the case of anti-loitering policies, but also for restaurants, bars, and pubs – disconnection is a way to improve customer experience. As the owner of HotBlack Coffee, a Toronto-based cafe with limited Wi-Fi service, explained to the press, 'It's not because we're trying to drive more business – people say, "Oh, you're just trying to turn more tables". That's not true. People sit here for hours. It's about an ambience and a social vibe' (Bouw, 2017). For some proprietors, it is also about food and drinks, which should be enjoyed and appreciated without the distraction of glaring screens and clacking keyboards (Dough Lover, 2020). In addition to taking away from the gustatory pleasures of dining out, digital devices are seen as a visual nuisance 'polluting a thoughtfully designed lighting scheme with the LED glow' (Barmann, 2017), an aesthetic threat and a safety and security hazard, as they can obstruct service pathways or be targets for theft (Raabe, 2014). But most of all, disconnection is supposed to relieve customers from the pressure of 24/7 connectivity by allowing them to 'relax and unplug' (Kibbitznest, 2020), 'slow down' (Hopkins, 2018), 'spend some quality time' (Barns, 2016), and 'live in the moment' (Gander, 2016). Resorting to such 'technology-free hygge havens' (Schwartz, 2019) often implies going back in time to an imagined earlier era when digital technologies were not all-pervasive – hence the popular trope 'pretend it's the 90s' and the frequent use of the words 'old-fashioned', 'old-school', and 'throwback'. To cafes, disconnection also means protecting their traditional role as a place of leisure rather than work and orchestrating the atmosphere accordingly to prevent 'the laptop brigade' from 'killing the mood' for other customers (Famurewa, 2016). Although bar and pub owners are usually spared this trouble, they are equally interested in creating a lively ambience in their venues. To do so, they ban mobile phones, encouraging customers to 'stop looking at their screens and talk to each other instead' (BBC, 2016).

Regardless of the type of business and the targeted technology, third place proprietors tend to consider unmediated communication to be a 'true', 'real', 'actual', and 'genuine' form of social interaction that requires restoration or cultivation. Hence, disconnection is often presented as their aspiration to bring back the allegedly lost 'art of conviviality' (Gander, 2016; Herd, 2018). Even though the concept of art implies making a creative effort, there is a persistent assumption that the sought-after 'social vibe' can be achieved by simply obstructing, restricting or prohibiting the use of digital technologies. To quote the owner of August First bakery in Burlington, US, 'We've noticed ... a liveliness to the bakery that wasn't there before. When people are on their laptops, they're silent ... Now it feels more alive, because of all the conversation' (De Seiffe, 2014). Thus, any expected or actual changes in customer behaviour after introducing a disconnection policy are usually attributed to the lack of connectivity rather than the exclusion of laptop workers and solo guests or other factors that might be at play, such as constructive socio-spatial tactics meant to encourage conversation and evoke a nostalgia for the analogue past.

For instance, Fox in the Snow cafe in Columbus (US) lowered the counter and removed pastry labels and prices to spark conversations between customers and baristas (Metz, 2017). HotBlack Coffee in Toronto, in addition to hiring people with good interpersonal skills, bought 'very low-profile' coffee machines, '[which] means you can look someone in the eye when you're making them their drink and carry on [the] personal contact' (CBC, 2017). Such seemingly phatic interactions with staff may also serve to draw customer's attention to the quality of food and drinks offered in the venue. Four Barrell Coffee in San Francisco offers books and magazines, plays music on a vinyl record player, and roasts beans on a 1950s coffee roaster (Guynn, 2010). Birch Coffee in New York designed conversation-starter cards ('Tell me about the one life-changing book I should read') that customers can place on their tables to invite strangers to have a chat (Kadet, 2017). In Democracy cafe in Hamilton, Canada, every wall is a blackboard where customers can write 'tweets' in chalk (Hamilton Spectator, 2013). Common tactics used in disconnected pubs include lowering the music volume (or not playing any music at all) and choosing not to install TV screens and gambling machines.[2] Across all types of businesses, disconnected venues tend to be decorated with natural, worn, and reclaimed materials and textures.

On the one hand, disconnection policies aimed at enhancing customer experience seem to benefit both sides of the market. They strive to provide customers with a sense of comfort, rootedness, belonging, and connection (not connectivity), while allowing small independent proprietors to showcase what differentiates them from large corporate chains – handcrafted food and drinks, unique design, and a community feel. On the other hand, this promise of community should be taken with a pinch of salt for it rests on three questionable assumptions. First, as previously mentioned, such policies tend to draw a causal link between disconnection and increased sociability, whereas existing ethnographic research has not revealed a deterministic relation between the presence of digital technologies and the number of face-to-face interactions in third places (Hampton and Gupta, 2008). Second, the discourse on disconnected third places often confuses bonding social ties within already existing family and friend groups with bridging ones, arising from social encounters between strangers (Putnam, 2000). That is, the lively buzz of conversation, commonly envisioned as the desired outcome of disconnection, does not necessarily mean that new contacts are being made – unless some constructive tactics are put into play.

Third, disconnection policies draw on the normative idea of third place as a space where 'conversation is the main activity' (Oldenburg, 1989: 49), which does not always hold up to historical scrutiny (Cowan, 2005) and fails to acknowledge the value of public solitude and inattentive forms of sociality for hospitality customers (Jones et al, 2015; McCamley and Morland, 2021). Moreover, advocates for disconnected third places tend to imagine the relationship between conviviality and digital technology as mutually exclusive, which does not acknowledge the hybridity of physical and digital spaces, practices, and experiences in today's technology-saturated world (Ridell and Zeller, 2013; Henriksen and Tjora, 2018). Not only can digital devices be woven into conversations between friends and family members (for example by showing each other photographs or looking up information online), but they can also provoke brief yet meaningful interactions between strangers, who often help each other find the Wi-Fi password, fix a software glitch, Shazam a song or guard one another's devices during bathroom visits (Tjora, 2013; Kviat, 2020).

Brand positioning

Disconnection can do more than just improve customer experience – sometimes, it becomes the backbone of a business's entire brand concept, or an essential part of it. The Gin Tub bar in the city of Brighton and Hove (UK) is not known so much for its drinks as for its Faraday cage, built into the walls and ceiling to prevent customers from using their mobile phones and encourage conversations between strangers. To create an atmosphere of serendipity and fun, the owners also introduced a table-sharing policy and equipped every table with a vintage rotary dial phone that can be used for placing orders at the bar or calling strangers on other tables (BBC, 2016; Gander, 2016). Altogether, these creative solutions form a unique selling point setting the Gin Tub apart from its many competitors in Brighton and Hove's bustling nightlife scene. Ironically, it also distinguishes it from its downstairs neighbour Shimmer, a venue owned by the same entrepreneur and positioned as 'the world's first selfie bar' – a place embracing and celebrating social media culture.

As some other examples suggest, the Gin Tub's success may have been due to the fact that Brighton and Hove is a 'tech-loving [city] full of students and freelancers' (Argus, 2018). It seems that the 'luxury of being offline' (Bowles, 2019) is particularly (though not exclusively) appealing to residents of tech hubs – places producing and produced by connectivity culture (Guynn, 2010; Famurewa, 2016; Barmann, 2017).

However, disconnection works just as well in a wider cultural and economic context shaped by the globalization, rationalization, and homogenization of urban space, such as in the West End of London: 'It is inarguable that Dean Street has been gobbled up by Crossrail, Tesco Metro and bleak coworking spaces where the macchiato-fuelled shout into Skype about start-ups. This is why [the] French House is so welcome. Because here is a place to talk all afternoon about nothing' (Dent, 2018).

The French House, calling itself 'the most iconic of Soho watering holes [where] actors, writers, artists and wits rub shoulders with royalty, bohemians and the film world', uses disconnection to build an image of an edgy, artistic urban pub with more than 100 years of history: '[Our] no music, no [gambling] machines, no television and no mobile phones rule makes [us] a haven for conversationalists and a firm favourite among some of the best-known names in showbusiness' (The French House, 2020). Contrastingly, for the seventeenth-century Poltimore Arms pub, disconnection is part of its identity as an off-grid gem, hidden in a remote rural area of Devon County in southwest England. The pub, welcoming 'everyone from farmers to Formula 1 stars and billionaires', is not connected to mains electricity, water, and gas, and the landlord rides a horse instead of driving a car. Commenting on his decision not to offer Wi-Fi, television and gambling machines, he said: 'I go to the pub to get away from those things, and so do other people – they come here for a conversation'. To make conversations flow effortlessly, the landlord introduces every newcomer to the crowd and asks them to 'have a go serving behind the bar'. The Poltimore Arms' customers are also invited to play the pub's piano and guitar and share a weekly communal meal cooked by a local resident. Food orders, however, are made online through a WhatsApp group. What is more, the pub's cat, named Frederic Albert Hitler for his naughty personality, has a personal Facebook page (Cooper, 2020).

A very different approach to disconnection can be found in Samuel Smith Old Brewery (also known as Samuel Smith's), a UK company that has been run by the same family since the nineteenth century. Over the years, Samuel Smith's has bought over 200 historical pubs around the country and gained a controversial reputation for its strict, zero-tolerance policies banning all types of digital devices – along with music, television, gambling machines and swearing – to preserve its 'traditional, uncompromisingly Victorian aesthetic' (Greer, 2019). Customers reviewing Samuel Smith's pubs on Google Maps often complain about the excessively aggressive policing of the company's rules by the staff (one reviewer even wondered if 'they're actors [hired] to be grumpy barmaids as part of a theme'). In 2020, IT expert Sally Lait, lectured by a member of staff about 'the evils of digital addiction' during her visit to an unnamed Samuel Smith's pub, opened up a discussion 'about the dangers of privilege and lack of empathy when it comes to increasingly popular "digital free" experiences'. In a series of tweets that received thousands of likes, Sally gave a few examples of situations where 'a device is genuinely essential for someone', including those who are visually and/or hearing impaired, have autism or social anxiety, have to use healthcare apps or need help with a foreign language (Lait, 2020).

Even though Samuel Smith's case is an exception rather than the norm, it demonstrates how the fetishization of tradition and heritage, often built into branding narratives justifying the use of disconnection policies in third places, can lead to social exclusion. While the Poltimore Arms' landlord also chose to, in his own words, 'go backwards' (Cooper, 2020), he did so by reviving the historical ideal of the third place as a societal leveller (Oldenburg, 1989: 47) rather than by protecting the aesthetic purity of the 'traditional pub' imaginary. In the French House, disconnection is linked to a different type of privilege – one based on taste and class distinction. Speaking more broadly, even in those cases where digital withdrawal is not meant to display a business's heritage, customer perceptions of disconnected third places typically range from 'behind the times' to 'cool and hip'. Apparently, one needs to possess a certain kind of cultural capital (in addition to a certain level of economic security) to be able to recognize and appreciate the cool factor of being offline.

Promotion

Because human dependence on ICT has always been a hot topic, disconnected third places easily get press coverage on a local, national, and even international level. According to some proprietors, media attention can increase sales so quickly that this sudden boost can be mistaken for a direct effect of disconnection on customer behaviour (De Seiffe, 2014). However, such publicity is not always welcome or solicited, especially for those who do not rely on disconnection as a branding strategy. For instance, the owner of Dough Lover felt that her business was 'used to make a point' by journalists, who may have overreported the cafe's disconnected status while neglecting her team's gastronomic efforts (see the comments section in Argus, 2018).

Given that in a post-Web 2.0 world a great deal of promotion work is done by customers, most smartphone-free venues only ban texting, calling and scrolling, but allow taking pictures (BBC, 2016) and even remind patrons to post them on social media after they leave (Kibbitznest, 2020). Across all types of businesses, owners and staff usually try to avoid negative reviews by steering away from direct confrontations with those who break or dispute their rules. However, customer backlash is not uncommon, particularly in cases where disconnection policies are enforced too aggressively or not explained clearly. It seems that the best practice in this regard is to use proper signage, be honest and transparent about the purpose of disconnection, and communicate this decision in a friendly and cooperative manner (see Niezgoda, 2019).

Non-economic goals

As Dough Lover's owner aptly responded to a reader's comment accusing her of disguising her 'true' (that is, profit-driven) motives, 'No cafe opens because the owner is just out for profit. Our margins are too small for that' (Argus, 2018). Indeed, for many proprietors, running a small independent business means more than just making profit – it is also their labour of love, a form of artistic expression and a way to contribute to their local community. Hence, their pursuit of conviviality is essentially a pursuit of the ideal third place (vibrant and thriving, filled with chatter and laughter) and the ideal self in it (a person rather than a function, a 'public character' (Oldenburg, 1989: 18) who knows everybody in the neighbourhood):

> According to [Australian cafe owners] Ostrick and Harrison, cultivating a close community outside the context of devices supports staff as much as providing a more social environment for those they serve. 'We have a really tight team because my staff feel part of something', Harrison reflects. Not just there to do the coffee and the food, they become central to the social network. They become human. … 'I've been invited to a 40th birthday, to a wedding anniversary dinner; I've had surfboards shaped for me', says Ostrick, recalling the multitude of important people in her life who began as regular customers to her cafe. (Lewis, 2017)

Furthermore, disconnected venues often reflect their founder's personality, background and ambitions. The owner of Télescope cafe in Paris calls himself a 'classical' person who prefers to have lunch in a bistro and work in an office, not the other way round. For him, disconnection is a way to 'go back to different values than the fast ones of Google, Wi-Fi and phones' (Hopkins, 2018). While some entrepreneurs find inspiration in the past, others draw it from their experiences of foreign cultures, including European, Latin American, and East Asian, all perceived as a romanticized ideal of laid-back, unconstrained sociability to which to aspire (Barns, 2016; CBC, 2017).

The Chicago bookbar Kibbitznest, founded by real estate developers and philanthropists Anne and Lewis Kostiner, grew out of their non-profit organizations aiming to 'preserve the quality of human interaction by promoting a balance between face-to-face and electronic communication media' (Kibbitznest, 2020). Inspired by 'the intellectually stimulating literary salons of days past', Kibbitznest was envisioned as a cultural intervention into the present and future:

> We have to create the culture because if we expect people to just change on their own without anyone taking action, it might not happen. Our responsibility is to extend ourselves beyond the electronic support systems that we have that are so easy. It's important because one of the basic aspects of being a human being is to communicate with each other. (Anne Kostiner in Bloom, 2016)

To achieve this mission, the Kostiners built a multisensory analogue sanctuary where there is no Wi-Fi, laptops and tablets are not allowed, smartphone use is 'strongly discouraged', walls are lined with books, newspapers, magazines, board games, and theme art (for example, two birds saying to each other, 'Let's tweet!' – 'No, let's talk'), seats are soft and cosy, customers can listen to vinyl records or the radio, watch old films on television screens (suddenly not so analogue), and write poems and notes on typewriters. In addition, Kibbitznest offers an extensive programme of cultural events.

Such initiatives, along with the growing popularity of the wider disconnection culture, do not go unnoticed by corporate chains. For instance, the UK's Beefeater and Frankie & Benny's experimented with 'no phone zone' campaigns encouraging families to spend time together away from screens. However, these were merely one-off promotional activities. For the time being, disconnection as a social and cultural mission is largely the preserve of small independent businesses.

Conclusion

Amid the increasing digitization of everyday life, a growing number of cafes, restaurants, pubs, and bars choose to go against the grain and limit or fully ban the use of digital technologies on their premises. Praised by many for their promise to 'bring back the art of conviviality', disconnected third places are just as often suspected of disguising profit-seeking manoeuvres as goodwill. However, a closer analysis of their historical and theoretical background and practical implementation reveals that social and economic aspects of digital disconnection in the hospitality industry should not be seen through the either/or lens. For instance, one of the most popular disconnection policies, the laptop ban, can be used to foster face-to-face interaction between customers, relieve them from the pressure to be 'always on' and thus contribute towards a healthier and happier society. At the same time, it can increase sales by speeding up table turnover, enhancing customer experience and facilitating market positioning and promotion. That said, economically motivated disconnection is not necessarily driven by capitalist greed – if anything, it helps small independent proprietors compete against large corporate chains. Furthermore, disconnection policies may reflect entrepreneurs' personal backgrounds and creative ambitions and make them feel part of their local communities.

Nevertheless, the discourse on disconnected third places rests on several questionable but persistent assumptions and imaginaries shared by both sides of the market and fuelled by the media. Not only does it perpetuate the long-standing tradition of viewing the relationship between unmediated and mediated communication as a binary opposition between sociable and unsociable, healthy and unhealthy, authentic and inauthentic, real and illusory, but it also tends to idealize the past while dismissing the present. Despite making an attempt to resist the ongoing colonization of leisure by work, disconnection policies may also feed into the neoliberal idea of personal responsibility and exclude and stigmatize individuals and activities not fitting the normative idea of who and what third places are meant for (not to mention the cultural and economic capital one needs to appreciate and afford the luxury of being disconnected). Drawing on the principles of empathy and non-discrimination, communicating with customers in a clear, honest, and cooperative way, and supporting obstructive, restrictive, or prohibitive policies with constructive ones is the best way to disconnect a third place, be it motivated by social or economic reasons, or both.

Ultimately, the rise of disconnected third places reflects the increasing pressure on small hospitality businesses to remain profitable while compensating for the lack of social services and infrastructures under neoliberal governance. It is yet to be seen whether the disconnection trend will increase or decline in the wake of the COVID-19 pandemic and the global cost of living crisis.

Notes

1. Another popular version of the 'We Don't Have Wi-Fi, Talk to Each Other' sign ends with the words 'Pretend It's 1995' (Ewen, 2020), when the internet (Web 1.0) was present but not yet ubiquitous.
2. Not having music and television in a pub also means no performance licence fees.

Acknowledgement and funding

This work was supported by the Economic and Social Research Council (Grant No.: ES/T008814/1).

References

Anderson, N. (2006) 'Free WiFi Spawns Cafe Backlash', *Ars Technica*, [online] 7 October, Available from: https://arstechnica.com/uncategorized/2006/07/7226-2/ [Accessed 12 February 2023].

Argus (2018) 'Dough Lover Cafe Bans Laptops and Tablets to Encourage Conversation', *The Argus*, [online] 19 February, Available from: https://www.theargus.co.uk/news/16031759.dough-lover-cafe-bans-laptops-tablets-encourage-conversation/ [Accessed 2 March 2023].

Barmann, J. (2017) 'Laptops in Restaurants: Nuisance, or Menace?', *SFist*, [online] 21 February, Available from: https://sfist.com/2017/02/21/etiquette_question_laptops_in_resta/ [Accessed 18 March 2023].

Barns, S. (2016) 'No Phones for You! A Japanese Restaurant Launches a Strict Mobile-free Policy', *The Daily Mail*, [online] 18 January, Available from: https://www.dailymail.co.uk/femail/food/article-3404712/Japanese-restaurant-Auradaze-launches-strict-mobile-free-policy.html [Accessed 18 February 2023].

BBC (2016) 'Hove Bar Uses Faraday Cage to Block Mobile Phone Signals', *BBC News*, [online] 3 August, Available from: https://www.bbc.co.uk/news/uk-england-sussex-36943686 [Accessed 19 February 2023].

Bloom, M. (2016) 'Laptops Banned at New Cafe/Book Store Opening in Lincoln Park', *DNA Info*, [online] 9 June, Available from: https://www.dnainfo.com/chicago/20160609/lincoln-park/laptops-banned-at-new-cafebook-store-opening-lincoln-park/ [Accessed 18 February 2023].

Bouw, B. (2017) '"We Don't Want to Be an Office": Café Owners Are Pulling the Plug on WiFi', *The Globe and Mail*, [online] 30 April, Available from: https://www.theglobeandmail.com/report-on-business/small-business/sb-managing/we-dont-want-to-be-an-office-cafe-owners-are-pulling-the-plug-on-wifi/article34859828/ [Accessed 23 March 2023].

Bowles, N. (2019) 'Human Contact Is Now a Luxury Good', *The New York Times*, [online] 23 March, Available from: https://www.nytimes.com/2019/03/23/sunday-review/human-contact-luxury-screens.html [Accessed 7 February 2023].

Boyer, M.C. (1992) 'The Imaginary Real World of CyberCities', *Assemblage*, 18: 114–127.

CBC (2017) 'Pulling the Plug', *CBC*, [online] 2 May, Available from: https://www.cbc.ca/player/play/934120515924 [Accessed 18 March 2023].

Cooper, J. (2020) 'Inside the "Off-grid" Pub Where Customers Serve the Drinks and the Cat's Called Hitler', *Devon Live*, [online] 15 February, Available from: https://www.devonlive.com/news/devon-news/inside-off-grid-pub-customers-3839733 [Accessed 12 March 2023].

Cowan, B. (2005) *The Social Life of Coffee: The Emergence of the British Coffeehouse*. Yale University Press.

Dawson, H. (2019) 'Coffee Shop Owner Is Accused of Bullying', *The Daily Mail*, [online] 10 November, Available from: https://www.dailymail.co.uk/news/article-7671033/Coffee-shop-owner-sparks-anger-BANNING-laptops-cafe-weekends.html [Accessed 3 March 2023].

Dent, G. (2018) 'The French House, London: "Stepping Out of a Time Machine" – Restaurant Review', *The Guardian*, [online] 16 November, Available from: https://www.theguardian.com/food/2018/nov/16/grace-dent-french-house-soho-london-restaurant-review [Accessed 22 February 2023].

De Seife, E. (2014) 'Burlington Bakery August First Goes Screen Free', *Seven Days*, [online] 16 April, Available from: https://www.sevendaysvt.com/vermont/burlington-bakery-august-first-goes-screen-free/Content?oid=2351162 [Accessed 21 February 2023].

Di Marino, M. and Lapintie, K. (2020) 'Exploring Multi-Local Working: Challenges and Opportunities for Contemporary Cities', *International Planning Studies*, 25(2): 129–149.

Dolan, D. and Daniel, J. (2009) 'The Daily Grind', *The New York Times*, [online] 27 November, Available from: https://archive.nytimes.com/www.nytimes.com/interactive/2009/11/27/opinion/28opart.html?pagewanted=all [Accessed 18 February 2023].

Dough Lover (2020) 'What Makes Us Different?' *Dough Lover*, [online] Available from: https://www.doughlover.com/about [Accessed 10 March 2023].

Ewen, N. (2020) '"Talk to Each Other Like It's 1995": Mapping Nostalgia for the 1990s in Contemporary Media Culture', *Television & New Media*, 21(6): 574–580.

Famurewa, J. (2016) 'Is the Jig Finally Up for Coffee Shop Wi-Fi Freeloaders?', *The Guardian*, [online] 20 November, Available from: https://www.theguardian.com/lifeandstyle/shortcuts/2016/nov/20/is-the-jig-finally-up-for-coffee-shop-wi-fi-freeloaders [Accessed 3 March 2023].

Fleishman, G. (2005) 'Some Cafe Owners Pull the Plug on Lingering Wi-Fi Users', *The New York Times*, [online] 13 June, Available from: https://www.nytimes.com/2005/06/13/technology/some-cafe-owners-pull-the-plug-on-lingering-wifi-users.html [Accessed 2 March 2023].

Ferreira, J. (2017) 'Café nation? Exploring the Growth of the UK Café Industry', *Area*, 49(1): 69–76.

Gander, K. (2016) 'Welcome to the Gin Tub, the Bar Encased in Copper to Stop You Looking at Your Phone', *The Independent*, [online] 9 August, Available from: https://www.independent.co.uk/life-style/gadgets-and-tech/gin-tub-bar-brighton-review-no-phones-survival-drinks-a7178976.html [Accessed 3 March 2023].

Greer, S. (2019) 'Samuel Smith's Pubs Are Banning Customers from Using Mobile Phones, Laptops and Tablets', *The Manchester Evening News*, [online] 29 March, Available from: https://www.manchestereveningnews.co.uk/news/greater-manchester-news/samuel-smiths-pubs-banning-customers-16046598 [Accessed 3 March 2023].

Gumpert, G. and Drucker, S.J. (2005) 'Communication and Urban Life', *Intermedia*, 33(2): 18–23.

Guynn, J. (2010) 'Coffee Shops Are Taking Wi-Fi Off the Menu', *Los Angeles Times*, [online] 8 August, Available from: https://www.latimes.com/archives/la-xpm-2010-aug-08-la-fi-cafe-wifi-20100808-story.html [Accessed 8 February 2023].

Hahn, F. (2004) 'The Saloon: A True Chat Room', *The Washington Post*, [online] 28 August, Available from: https://www.washingtonpost.com/archive/lifestyle/2004/08/27/the-saloon-a-true-chat-room/a95541c1-af62-40e4-a334-db24aa337267/ [Accessed 3 February 2023].

Hamilton Spectator (2013) 'Locke Street's New Coffee House Encourages Conversation – Even Political Discourse', *Hamilton Spectator*, [online] 11 September, Available from: https://www.thespec.com/business/2013/09/11/locke-street-s-new-coffee-house-encourages-conversation-even-political-discourse.html?rf [Accessed 18 February 2023].

Hampton, K. and Gupta, N. (2008) 'Community and Social Interaction in the Wireless City: Wi-Fi Use in Public and Semi-public Spaces', *New Media & Society*, 10(6): 831–850.

Hauser, A. (2018) 'La Colombe Coffee Breaks No-Internet Policy, Starts Offering Customers WiFi', *Eater Chicago*, [online] 2 January, Available from: https://chicago.eater.com/2018/1/2/16842602/la-colombe-internet-wifi-offer-weekday [Accessed 18 March 2023].

Henriksen, I. and Tjora, A. (2018) 'Situational Domestication and the Origin of the Cafe Worker Species', *Sociology*, 52(2): 351–366.

Herd, M. (2018) 'No Laptops Allowed – The Cafes Bringing Back the Art of Conviviality', *The Guardian*, [online] 12 February, Available from: https://www.theguardian.com/society/2018/feb/12/no-laptops-allowed-the-cafes-bringing-back-the-art-of-conviviality [Accessed 3 February 2023].

Hickman, P. (2013) '"Third Places" and Social Interaction in Deprived Neighbourhoods in Great Britain', *Journal of Housing and the Built Environment*, 28(2): 221–236.

Hopkins, R. (2018) 'Nicolas Clerc on "Télescope", a Café Without Smartphones: "They're Filling Our Lives with Emptiness"', *Rob Hopkins*, [online] 23 November, Available from: https://www.robhopkins.net/2018/11/23/nicolas-clerc-on-teescope-a-cafe-without-smartphones-theyre-filling-our-lives-with-emptiness/ [Accessed 3 March 2023].

Hubbard, P. (2019) 'Enthusiasm, Craft and Authenticity on the High Street: Micropubs as "Community Fixers"', *Social & Cultural Geography*, 20(6): 763–784.

Jeffres, L., Bracken, C., Jian, G., and Casey, M. (2009) 'The Impact of Third Places on Community Quality of Life', *Applied Research in Quality of Life*, 4: 333–345.

Jones, H., Neal, S., Mohan, G., Connell, K., Cochrane, A., and Bennett, K. (2015) 'Urban Multiculture and Everyday Encounters in Semi-public, Franchised Cafe Spaces', *Sociological Review*, 63(3): 644–661.

Kadet, A. (2017) 'What Happens When Cafes Pull the Plug on Wi-Fi', *The Wall Street Journal*, [online] 19 January, Available from: https://www.wsj.com/articles/what-happens-when-cafes-pull-the-plug-on-wi-fi-1484669953?mod=e2tw [Accessed 12 March 2023].

Kelly, T. (2009) 'Pub Where You Are Barred from Using Your Mobile Phone', *The Daily Mail*, [online] 2 September, Available from: https://www.dailymail.co.uk/news/article-1210601/Pub-barred-using-mobile-phone.html [Accessed 12 March 2023].

Kibbitznest (2020) 'About Us', *kibbitznest*, [online] Available from: http://www.kibbitznest.com/about [Accessed 1 March 2023].

Kuntsman, A. and Miyake, E. (2016) 'Paradoxes of Digital Dis/engagement: A Follow up Study (Businesses and Services)', *Working Papers of the Communities & Culture Network+*, 7: 1–16.

Kuntsman, A. and Miyake, E. (2019) 'The Paradox and Continuum of Digital Disengagement: Denaturalising Digital Sociality and Technological Connectivity', *Media, Culture & Society*, 41(6): 901–913.

Kviat, A. (2020) 'Connected and Sociable: Third Places in the Post-digital City', *CAMEo Cuts*, 14, Available from: https://le.ac.uk/cameo/publications/cameo-cuts [Accessed 23 March 2023].

Lait, S. (2020) *Twitter*, [online] 5 January, Available from: https://twitter.com/sallylait/status/1213799200498077696 [Accessed 3 March 2023].

Lewis, S. (2003) 'The Integration of Paid Work and the Rest of Life: Is Post-industrial Work the New Leisure?', *Leisure Studies*, 22(4): 343–345.

Lewis, S. (2017) 'Café Owners Are Cultivating Community by Saying No to Wi-Fi', *SBS*, [online] 25 May, Available from: https://www.sbs.com.au/food/article/2017/05/25/how-would-you-feel-about-no-wi-fi-your-fave-cafe [Accessed 12 March 2025].

Lugosi, P. (2007) 'Consumer Participation in Commercial Hospitality', *International Journal of Culture, Tourism and Hospitality Research*, 1: 227–236.

McCamley, C. and Morland, L. (2021) 'Lone Not Lonely: Conceptualising the Lone Consumer Servicescape through Speciality Coffee', *Marketing Theory*, 21(3): 371–389.

Meier, R.L. (1962) *A Communications Theory of Urban Growth*. MIT Press.

Metz, E. (2017) 'Some Cafes Are Banning Wi-Fi to Encourage Conversation', *BBC*, [online] 27 February, Available from: https://www.bbc.com/worklife/article/20170224-some-cafes-are-banning-wi-fi-to-encourage-conversation [Accessed 27 March 2023].

Mimoun, L. and Gruen, A. (2021) 'Customer Work Practices and the Productive Third Place', *Journal of Service Research*, 24(4): 563–581.

Niezgoda, A. (2019) 'Brookline Café Ditches Wi-Fi After Business Turns into "Sea of Laptops"', *NBC Boston*, [online] 22 August, Available from: https://www.nbcboston.com/news/local/allium-market-brookline-massachusetts-says-goodbye-to-wifi/89393/ [Accessed 18 March 2023].

Oldenburg, R. (1989) *The Great Good Place: Cafes, Coffee Shops, Community Centers, Beauty Parlors, General Stores, Bars, Hangouts, and How They Get You Through the Day*. Paragon House.

Putnam, R.D. (2000) *Bowling Alone: The Collapse and Revival of American Community*. Simon & Schuster.

Raabe, S. (2014) 'Colorado Coffee Shops Putting Limits on Power, Wi-Fi Use', *The Denver Post*, [online] 4 July, Available from: https://www.denverpost.com/2013/07/04/colorado-coffee-shops-putting-limits-on-power-wi-fi-use/ [Accessed 17 March 2023].

Ridell, S. and Zeller, F. (2013) 'Mediated Urbanism: Navigating an Interdisciplinary Terrain', *International Communication Gazette*, 75(5–6): 437–451.

Schulte, S.R. (2011) 'Cutting the Cord and "Crying Socialist Wolf": Unwiring the Public and Producing the Third Place', in D.W. Park, N.N. Jankowski, and S. Jones (eds) *The Long History of New Media: Technology, Historiography, and Contextualizing*. Peter Lang, pp 37–54.

Schwartz, A. (2019) 'Open Tab: Just a Metra Ride Away, Kibbitznest Offers a Technology-free Hygge Haven', *The Daily Northwestern*, [online] 6 March, Available from: https://dailynorthwestern.com/2019/03/06/ae/open-tab-just-a-metra-ride-away-kibbitznest-offers-a-technology-free-hygge-haven/ [Accessed 23 February 2023].

Sennett, R. (1977) *The Fall of Public Man*. Knopf.

Stabiner, K. (2018) 'What to Do When Laptops and Silence Take Over Your Cafe?', *The New York Times*, [online] 13 February, Available from: https://www.nytimes.com/2018/02/13/dining/cafe-wifi-remote-workers.html [Accessed 18 February 2023].

Stäheli, U. and Stoltenberg, L. (2024) 'Digital Detox Tourism: Practices of Analogization', *New Media & Society*, 26(2): 1056–1073.

The French House (2020) *French House Soho*, [online] Available from: http://www.frenchhousesoho.com [Accessed 18 March 2023].

Tjora, A. (2013) 'Communal Awareness in the Urban Café', in A. Tjora and G. Scambler (eds) *Café Society*. Palgrave Macmillan, pp 103–126.

Turkle, S. (2011) *Alone Together: Why We Expect More from Technology and Less from Each Other*. Basic Books.

4

Digital Representations on the Appalachian Trail: Wilderness as a Post-Digital Experience

Dave McLaughlin

Introduction

In the last decade, as digital devices and social media have become ever more common features of daily life, voices from various walks of American life – academic (Odell, 2019), technological (Newton, 2018), corporate (Mele, 2017), and religious (Frost, 2022) – have articulated the virtues of a movement termed 'digital detox'. The ambitions of each of these thinkers is quite different: whereas writer and artist Jenny Odell encourages us to step back from social media to help critically judge and so confront capitalistic modes of production (Odell, 2019), corporate guru Levi Felix offered digital detox as a tool to become even *more* productive (Mele, 2017; Odell, 2019); meanwhile, Christian charity Defend Young Minds proposes digital detox as a tool to combat pornography use among teenagers (Frost, 2022). Despite this, these diverse voices have one thing in common: they offer wilderness as the antithesis of our addicted and distracted digital practices.

Wilderness as an escape from the busy-ness of everyday life has a long and deep resonance in non-indigenous American culture (and in Western culture more generally) (see Cronon, 1996; Odell, 2019). From John Muir (1992) to Levi Felix (Mele, 2017) via Benton Mackaye (see Minteer, 2001), Earl Shaffer (1983/1948), Bill Bryson (1998), and Rebecca Solnit (2002), Americans have represented and practiced wilderness environments as sites of escape from everyday busy-ness. Despite the complicated and problematic nature of this conception of 'the wild', it has retained its power as a site of escape in our contemporary age of digital saturation. The Appalachian Trail (affectionately known as the 'AT' to those who hike it), the oldest long-distance hiking trail in America (see Minteer, 2001) stands as a totem of the American wilderness experience (see Nash, 2014; Brown, 2019) – and is particularly popular among those looking to practice digital detox, of varieties both quick and restorative (see Fondren, 2016: 31; Thompson, 2020; Frost, 2022; McCadams, 2022; Glebe, 2025).

In this chapter, I set out on a path not of reinforcing the idea of the Appalachian Trail as a site for people to get away from the digital and back to their 'authentic' selves (termed in this collection the 'analogue idyll'), but of problematizing this idea. Despite the popular image of the Appalachian Trail as a place where one is forced away from technology (see Fondren, 2016: 31), the twenty-first century has seen increasing numbers of hikers, both long-distance 'thru-hikers' and shorter 'section hikers' and day hikers, bringing digital devices onto the trail. As I shall discuss here, hikers use these devices for a variety of reasons (see also McLaughlin, 2022, 2023): they post entries on websites such as *trailjournals.com* and *whiteblaze.net*; they post pictures and messages on social media sites such as Instagram and Facebook; they even use apps to navigate the trail itself. In this chapter I want to ask: what happens to the idea of wilderness in general and the AT in particular as a site of digital detox, as an analogue idyll, when AT hikers bring their phones into the woods? I argue that a focus on AT hikers' use of digital devices on the trail, particularly the digital representations they produce, can help us to see this trail in the twenty-first century as a post-digital space sitting in-between the dream of wilderness as an analogue idyll and the reactive, digital busy-ness of everyday life.

Wilderness and walking as escapes from the (digital) world

As a long-distance walking trail surrounded by a wilderness corridor made by fiat, the Appalachian Trail can be contextualized by two sites or practices of escape from the world: wilderness and long-distance or deliberate walking (I say 'deliberate' walking here to distinguish this practice from quotidian bipedal locomotion). The idea of wilderness as a site of escape from the human-built, technologically infused everyday world and a return to a sense of naturalness or authenticity has a long history in non-indigenous American culture. Henry David Thoreau famously declared 'wildness to be the preservation on the world' (Cronon, 1996: 9), while John Muir attempted to define American wilderness as wholly outside of this mortal earth, declaring of Yosemite Falls that: 'No description of Heaven that I ever heard or read seems half so fine' (Muir, 1992: 211).

The continued popularity of American nature writers such as John Muir and Edward Abbey arguably rests on their presentation of the *idea* of wilderness as a place antithetical to the digital. As Abram writes: digital '[t]echnological visions remain motivated by … a fear of our carnal embedment in a world ultimately beyond our control – by our terror of the very wildness that nourishes and sustains us (Abram, 2011: 69). For author and artist Jenny Odell, attending to the continued existence of wild landscapes helps her to find escape from 'the effects of current social media … and its deliberately addictive features' (Odell, 2019: 15). Writing of an ancient, gnarled redwood known as Old Survivor, clinging on in the ravaged hills above Oakland, California, Odell notes that it is one example of ways of being which 'exceeds algorithmic description' (2019: 23) and which exemplifies a kind of 'resistance-in-place' to the ever-encroaching productivity engine driven by digital devices and social media (2019: 22). Going further, she explains that: 'To look at that tree is to look at something that began growing in the midst of a very different, even unrecognisable world … This rootedness is something we desperately need when we find ourselves awash in an amnesiac present and the chain-store aesthetic of the virtual' (Odell, 2019: 25).

A similar emphasis on specific parts of digital technologies, particularly social media, can be found in perhaps the most famous wilderness digital detox event in the world – Camp Grounded. The aesthetic of this 'summer camp for adults' (Digital Detox, 2023) emphasizes hallmarks of the American wilderness that John Muir would heartedly endorse: its logo shows a sunrise over a large redwood tree; its promotional material features camp-goers sitting in the woods and around fires; its employees dress in old-fashioned outdoor gear (Digital Detox, 2023). Yet the camp does not reject all technologies: activities rely on older technologies such as typewriters; camp amenities and lights run on electricity; and the event is organized and promoted via the internet (Odell, 2019: 93–94; Digital Detox, 2023). Camp Grounded's digital detox was aimed at social media and its activities, as some of their rules attest: 'No Digital Technology; No Networking; No Phones, Internet or Screens' (Odell, 2019: 94).

If knowing that wilderness is 'out there' can be a comfort for those looking to disengage from the digital, then getting out into the wild can be a means of practicing detachment. Such is the reason that long-distance walking maintains its popularity as a means of escaping from the busy-ness of everyday digital life. In his meditative textual excursion on the philosophies of walking, Frédéric Gros argues that long-distance walking is, above all, an exercise in freedom. It is, he writes, freedom from 'the profusion of what is available (merchandise, transport, networking), and easy access to facilities (to communicate, to buy, to move about)' (Gros, 2015: 4). What comes in its place is a life at once more constrained, yet freer due to its connections to a way of life more essential, even more animal, than we might feel in modern-day, networked, connected lives: 'tiresome constraints make themselves felt [when walking]: the rucksack, the length of the stages, the weather (threat of rain, of storms, of murderous heat), the primitive accommodation … When fog shrouds the mountains or rain starts to fall you have to continue, to keep going' (Gros, 2015: 3–4).

These apparent discomforts are not so, says Gros, for a very good reason: 'When walking in this mode we discover … elemental energies and our appetites follow: they are enormous and our bodies are satisfied' (Gros, 2015: 6). While Gros does not mention escape from the digital per se, his emphasis on the networks and connections of everyday life does raise it as a spectre haunting the world he wishes to escape. Gros's emphasis on a body satisfied in a more challenging, elemental environment also speaks to debates around the increasing distance between the dreams of digital-utopian visionaries such as Peter Thiel and Ray Kurzweil and the messy, fleshy realities of human bodies previously emphasized by Abram. He also alights on an idea noticed by another student of walking, Robert Moor, who has argued that 'the continued interest in hiking seems to stem from a desire to cut through the techscape to get to some natural substratum: to borrow [AT founder Benton] MacKaye's phrase, to see the "primeval influence" beneath the "machine influence"' (2016b: 261).

For others, long-distance walking harks back not to our animal selves but to a more historical manifestation of humanity embedded in and, perhaps, nostalgic for a pre-digital world. For anthropologist Tim Ingold, walking is seen as a defining characteristic of humans as embodied beings. If the practices of urban modernity leave us disconnected from the world, argues Ingold, walking can help restore us to it – both metaphorically and literally (Ingold, 2006: 24). As John Wylie notes, for Ingold, walking is less a matter of locomotion and more a matter of making oneself at home in the world, where 'wayfaring is a matter of ongoing craft and creation' (Wylie, 2016: 412; see also Ingold, 2010). A second manifestation of this idea takes the form of walking as pilgrimage. Sylvain Tesson's three-month voyage along long-forgotten paths of rural France, for example, in *The Wandering Paths* is framed explicitly in terms of both secular pilgrimage and digital disconnect (Tesson, 2022). Geographers researching modern pilgrimage similarly emphasize the idea that walking works to bring its practitioners back to a more authentic speed of life (see Maddrell and Della Dora, 2013; Scriven, 2021: 65–66). Pilgrims' 'interactions with the natural environment combine with the intentional mindfulness and slow pace of pilgrimages' to generate a sense of 'liminality … the detached nature of the journey's participants enter a transitional state enabling more meaningful experiences' (Scriven, 2021: 66). A key part of the experience of walking as pilgrimage, in this contemporary age defined by constant digital connections and the anxieties these often bring, reflecting the timbre of Jenny Odell's anti-capitalist call for disconnection, is to allow walkers to entertain alternative and more hopeful imaginations of themselves and of the societies in which they live. 'Solidarities formed on the route', writes Scriven, 'can generate deep bonds, and chance meetings can offer *reassurances* concerning human nature' (Scriven, 2021: 72, emphasis added).

Trail of technology

Despite its reputation as a popular site for digital detoxing, the Appalachian Trail is not quite the analogue idyll that listicle writers make it seem. Even though wilderness lovers and hikers have long cherished an image of the outdoors divorced from human influence, hikers are less exclusionary than they are ambivalent about their own use of technology in hiking experiences (see Syvertsen, 2022). Of the Appalachian Trail, Robert Moor writes that 'ironically, the act of hiking is also dependent on technology … Today, some forms of technology (like cell phones or ATVs) are considered obnoxious, while others (like water purifiers, camp stoves, and GPS locators) are excused. In either case, technology inexorably trickles into the wild' (Moor, 2016b: 261).

Research suggests that modern digital technologies, from GoPro cameras to GPS-enabled digital watches, are increasingly popular among hikers as tools to increase their physical performance and so heighten their experiences (see Cory, 2015; Stelter and McCrickard, 2017; Fondren and Brinkman, 2019). There is also evidence, despite the strength of Moor's earlier assertion, that smartphones are popular and useful, even for hikers on a digital detox. Smartphones, with their ability to make phone calls and send text messages, for example, have been found to increase senses of place-attachment on the Pacific Crest Trail (Amerson et al, 2020: 308–309; Syvertsen, 2022). On the Appalachian Trail, the most popular map of the route (displacing the old guidebook by Dan 'Wingfoot' Jones) is David 'AWOL' Miller's *The A.T. Guide*, which is only available as a smartphone app (see A.T. Guide, 2023). Finally, even Christian anti-pornography campaigner Lisa Frost's efforts at a family-friendly digital detox on the Appalachian Trail involve the use of 'a messaging app and an app to lead us to spring water and shelter' (Frost, 2022).

Over the past 20 years many AT hikers have embraced the social elements of digital technologies which seem, within the first flush of the digital detox movement, to be so antithetical to wilderness experiences. Far from identifying social media as a divisive force among hikers, distancing them from the landscapes through which they move and from the communities created on interpersonal connection, a small amount of recent scholarship demonstrates the community and identity-forging power of social media platforms (see Marx, 2019; McLaughlin, 2022, 2023). Both Phiona Stanley (2020) and Weatherby and Vidon (2018) demonstrate the power of social media platforms to increase a sense of belonging among groups traditionally excluded from American wilderness spaces: women, people of colour, disabled people, parents,.fat people and so on. This new research, coupled with the fact that at least 96% of hikers on American trails carry smartphones (Amerson et al, 2020), and the small but increasingly important pool of hikers making and sharing digital content (see McLaughlin, 2022), raises an important question: far from being an analogue idyll *par excellence*, can attending to hikers' digital representations including blogs, eBooks, online articles, and social media help us to understand these woods as a post-digital space?

'Digilogue' material culture

Attending to AT hikers' digital writings and representations, including blogs, Instagram posts, online articles and ebooks, reveals a world in which digital and non-digital material cultures mix together. Different accounts by twenty-first century hikers, some hiking 20 years ago, others more recently, illustrate the longevity of hikers switching back and forth easily between analogue and digital technologies as they walk along the trail. Kristi Fondren's scholarly account of her own hike on the AT and her encounters with other members of the AT community, made in the first half of the 2010s, bears this out. Despite being written and published in the age of the smartphone, Fondren's book gives a general impression of the AT hiking world as one entirely separated from the influences of digital social media which have such a hold on the outside world. In her book, words to describe the digital world hardly appear: 'phone' only three times; 'online' twice; 'digital' not at all. This is reflected in her encounters with hikers, such as Bonzo, who proclaim their wilderness experience as a deliberate act of disconnection from the world in explicitly non-digital terms: 'on the trail "you don't have the influence of radio, television, newspapers"' (Fondren, 2016: 31).

Yet, for all this, the Appalachian Trail that Fondren describes is not wholly non-digital. In an interview with Fondren after leaving the AT, a hiker using the trail name Turbo (a pseudonym used during their hike, like 'Bonzo') tells a story about the trust which is extended between hikers and non-hikers in the region of the trail. At the centre of Turbo's story is a computer: 'There was a sign [at Bear's Den Gap in Virginia] that said thru-hikers go around back … you go in and there's a free computer set up with Internet access for hikers, and a phone with free long-distance' (Fondren, 2016: 61).

While the Appalachian Trail that Fondren finds is more 'analogue' than 'digital', Turbo's story indicates that there is a place for the digital, too, even if a limited one. David Miller's account of hiking the AT in 2003, using the trail name AWOL provides a much more vivid image of a trail experience in which the digital has a clear place alongside the non-digital in hikers' lives. His account contains repeated instances of using computers to 'surf the

Internet' (Miller, 2010: 111): in the homes of trail angels, people who live the near the AT and provide assistance to hikers for free (2010: 111); in the Appalachian Trail Conservancy head office in Harper's Ferry (2010: 117); at a makeshift hiking shelter in the 'clean and spacious basement of the St. Thomas Episcopal Church' in Vernon, New Jersey (2010: 137) as well as at various internet cafes in towns along the trail.

Miller's observations at the Side Track Cafe in Damascus, Virginia, demonstrate the attitude of certain AT hikers (including himself) to the internet and its availability along the route of the trail, even in 2003:

> Hikers sit at four of the machines [computers], eagerly checking e-mail, typing their own journal entries, and looking up journals of other hikers they have met on the trail. Even though pay phones are available in every town, I prefer to send email … Using e-mail, I still retain a degree of separation from the 'real world'. (Miller, 2010: 53)

While the digital ecosystem of 2003 in which Miller moved did not include smartphones, it is telling here that Miller adopted digital communications technologies in place of a more readily available analogue one (the telephone), in order not to escape the wilderness, but to stay within its bounds.

As an early and keen adopter of digital technologies on the Appalachian Trail, David Miller's eBook account reveals much about the ease with which he found himself switching between digital and non-digital technologies as he hiked north. We have already seen that Miller maintained a fondness for digital communication. His quote that follows further illustrates the extent to which hikers in the early twenty-first century were already creating online journals (blogs taking the form of diaries) as a means of recording their experiences and communicating with friends at home and in the woods. Yet, as an IT professional who would go on to found the online-only A.T. Guide (2023), Miller actively embraces a mixture of digital and non-digital media to both record his journey and communicate with others, on and off the trail. As much as 'the Internet' finds repeated mention in Miller's published account of his hike, there are many more mentions of 'trail registers': notebooks in which hikers write thoughts, draw images, and communicate with each other.

These notebooks, known as registers, are found all along the trail: 'At each shelter, and in most hostels, there is a shelter journal [meaning register], a notebook and pen usually found in a plastic bag' (Fondren, 2016: 86). While '[r]egisters away from shelters are rare' (Miller, 2010: 124), there are some exceptions, such as the famous register at Mount Springer, marking the start of a northbound hike (and where hikers really do register their hikes officially with the Appalachian Trail Conservancy, the organization which officially oversees the trail). Others include a register at the concession stand in Pine Furnace State Park, Pennsylvania, where thru-hikers traditionally complete the Half Gallon Challenge, which involves consuming an oversized portion of ice cream and writing about it in the journal provided (see McLaughlin, 2023). Sometimes, registers appear incongruously out in the open: 'There is a mailbox with "register" written on the side, oddly placed on this inconsequential peak' (Miller, 2010: 124).

In Miller's writing, these registers form an important and integral part of the material culture of the trail. They provide a source of entertainment on dark and cold nights (Miller, 2010: 71); provide spaces for commemoration of the victims of rare murders (2010: 68); hold warnings and sitings of bears (2010: 51); and provide a means for hikers to recognize and communicate with each other – even if they never meet in person (2010: 9) Hikers' use of pen-and-paper trail registers is so ubiquitous during Miller's hike that he claims it is more than a habit: it is a hallmark of the thru-hiker: 'In a display of *yet another thru-hiker mannerism*, Ziggy goes straight for the spiral-bound notebook that serves as a shelter register' (Miller, 2010: 8, emphasis added).

Miller's accounts of switching between digital and non-digital technologies suggest that the material culture of the Appalachian Trail in the twenty-first century befits the term 'digilogue'. 'Digilogue' (or 'Digilog', in its original, American English spelling) is a term coined by Eric Barry to describe the medium of modern music: 'rather than purely digital or analog, both the recording and the distribution of music media are both' (Barry, 2014). For Barry, 'digilogue' describes a situation in which producers and consumers of media have the freedom to range between analogue technologies (such as vinyl record players) and digital technologies (such as music streaming services) without any sense of dissonance or cultural reversion. While Barry discusses music, we can see the Appalachian hikers we have already met operating under a similar logic. In David Miller's account, hikers who spend their time habitually writing in non-digital trail registers just as eagerly go online whenever they can to write in their own online journals and communicate electronically with people on and off the trail (Miller, 2010: 8–9, 53). Digilogue does not just suggest that media producers and consumers move freely between different formats: the information can move just as easily. As Barry writes: 'CD and MP3 listeners are digilog when they listen to the vast oeuvre of music originally recorded in analog or transferred to analog in the process of production' (Barry, 2014). We can see a similar logic at work when considering Miller's account of his AT hike in its totality. His account was originally written as notes in a paper notebook; transferred to an electronic document which was emailed to his wife; printed and circulated in his local Florida newspaper; and finally written up and published as an eBook to be read on e-readers and (in my case) a smartphone reading app (see Miller, 2010: 16, 19, 60).

Alternative affordances and post-digital practices

Hikers actively practice a post-digital material culture on the Appalachian Trail for the simple reason that digital and non-digital media and the technologies which enable them offer different affordances, or purposes, to hikers at different times. In 2016 and 2017, for example, two articles about hiking the Appalachian Trail were published by the online-only news site *Buzzfeed.com*. While the accounts of Rahawa Haile, an Ethiopian-American woman hiking to increase the representation of black people in American wilderness spaces, and Robert Moor, a white American man hiking for his own self-exploration, differ in their approach, each has something similar to say about their relationship to analogue and digital technologies. For both hikers and authors digital media formed a noticeable part of their trail experience. Haile posts photographs of paperback books by black authors which she carries with her on the trail (as I discuss next; see also McLaughlin, 2023). Moor describes meeting other hikers who make use of social media to document the changes their body undergoes (Moor, 2016a). Yet, both accounts also foreground the use of (mostly) non-digital forms of communications technology, particularly the postal service. Rather than post letters to others, both Haile and Moor describe posting things to *themselves*; objects whose individual weight would collectively add too much to the weight they had to bear in their packs. For Moor, posting supplies, including 'chapbook[s] of poetry or a heftier paperback novel I had cut into slimmer volumes' (Moor, 2016b: 9) allowed him to walk more lightly and move more swiftly. For Haile, the decision was more fraught with a sense of selfhood sundered by time and distance: 'Where can I ship the contents of my blackness? How much of it can I permit myself to carry at a time?' (Haile, 2017).

If digilogue describes a sense of freedom in moving between analogue and digital media, the term 'post-digital' describes thinking which is at once beyond and in opposition to the ultra-cool, total digital/technological utopias of early internet and social media. It suggests an integration of 'analogue' and digital together into a more matter-of-fact assessment of the role of digital – companies, devices, media – in our everyday lives. Yet the 'post' also speaks of a 'continuation' of the digital – we have not moved beyond the digital, just beyond the early twenty-first century. As Florian Cramer describes it: 'The notion of "post-digital" primarily [is] an antidote to techno-Hegelianism' in which "digital" had long been synonymous with "progress"' (Cramer, 2015: 14). Many hikers we have met already seem to think in these terms – if more for reasons of practicality than ideology or philosophy. Just as ride-hailing apps such as Lyft and Uber provide only a simulacrum of digital excitement draped over the analogue reality of a ride in a piece of very nineteenth-century technology piloted by a fellow human being in a position of economic precarity, so Haile and Moor's use of the equally nineteenth-century postal service to ship their goods speaks to both a recognition of the limits of digital technologies in the wilderness and a recognition of the relevant affordances of non-digital technologies and media.

Trail registers have a long history as a tool for direct communication between hikers (see Swift, 2015). David 'AWOL' Miller, for example, tells of his using a shelter register to suggest a trail name for one hiker: 'I suspect I won't see Mike again, so I leave a note in the shelter register, suggesting the trail name "Merlin." I hope he sees it' (Miller, 2010: 21). Later, Mike confirms in a letter to Miller that he did receive AWOL's message, due to the register and its central role in AT hiker's social world: 'Rain and Shine read your suggestion in the Brown Fork shelter log, that I use "Merlin" as my trail name, and I liked it (a lot actually), so "Merlin" it is' (Miller, 2010, fn 9). Sometimes these relations appear to be more ephemeral than communicative, with the trail registers playing a more active role as a social partner, rather than a kind of mountain telegram. Fondren's account of her own Appalachian hike includes mention of a trail register where an unnamed hiker has posed a question for all those coming after them: 'Tell me a conspiracy theory' (Fondren, 2016: 86). While other hikers have written responses, including one which Fondren quotes at length, it is unlikely that the original question poser will return to read them. These paper notebooks offer, in a way, a serial-like story experience for hikers on the trail, one which is more akin to the postmodernist *If on a Winter's Night a Traveler …* by Italo Calvino (1981) than to any regularly serialized tale, such as the Sherlock Holmes short stories.

The power of register entries to tell disjointed stories is related by scholar and hiker Roger Sheffer, in his study of the 'maps of the imagination' left behind by hikers in a selection of trail registers kept by hiking clubs up and down the Appalachians. Of one such memorable example, he writes:

> An anonymous artist has provided an illustrated map of 'the trail between here and Route 309', making it easier to take [this suggested but necessary bypass of the trail's usual route], perhaps, by turning various features of the landscape into familiar household objects: Bake Oven Knob becomes a four-burner stove with its door open, Bayer's Rocks are three big aspirin tablets set side-by-side, and, of course, the Knife Edge [a section of hill-top path] is a knife edge. (Sheffer, 2004: 66)

Other imaginative cartographers use hand-drawn mappings to emphasize the difficulty of steep ascents or descents already taken, the 'ant-like people' rendered tiny in comparison to the hills; or to make a weary statement about the sheer number of hills that hikers need to cover before the end (Sheffer, 2004: 66–67). These maps and notes arguably hold an affordance that digital media cannot. They are more than the direct, peer-to-peer guidance on the physical, emotional, and imaginative hardships of the trail, which is also available by reading hikers' online blogs. They are also, by virtue of being held in registers found in shelters along the trail, tangible, holdable signs that others have been this way before and survived to tell the tale – with all the wit and honesty they can muster.

A significant example of the differing affordances of digital and non-digital textual representations of hiking the AT can be found in the differing accounts of preparing to hike the trail offered by two of the most widely read hiker/authors of the past quarter-century: Bill Bryson and David 'AWOL' Miller. In *A Walk in the Woods* (1998), part nature writing, part hiking memoir, Bryson paints a picture of preparations for his Appalachian adventure which emphases the increasing isolation and quiet fear he felt as the days tick down to his official setting off. From reading an academic text about bear attacks, to feeling like a fish out of water in the camping goods store, to setting up his tent in the basement while his wife sleeps upstairs, Bryson's pre-trail activities are filled with a sense of apartness from his regular life and the people in it: the enactment of a gradual withdrawal from society (see Bryson, 1998).

Compare this to David Miller's account of his own pre-trail preparations, which come half-way through his tale, prompted by the 'happy reminder that I am on a grand adventure' (Miller, 2010: 161). Miller reminisces about reading along with the online postings of a hiker called Bono, who began hiking the AT on 1 January 2003: 'I eagerly followed his journal for four months before leaving home … Each new entry of Bono's journal exposed me to new sections of the trail, new places I longed to see for myself, and the emotional ups and downs of life on the trail' (Miller, 2010: 131–132).

Like Bryson, Miller prepares for life on the trail by sitting at home, alone. Yet unlike Bryson's analogue preparations, the affordances of Bono's online journal offer Miller a sense of companionship, belonging, and the excitement of entry into a new world which is very different from Bryson's steady uncoupling from his old self. Where printed books about bear attacks bring fear, online journals offer hope.

A key marker of belonging in the AT hiking community is the adoption of a trail name. Since the 1980s, it has become common practice among Appalachian Trail hikers to adopt trail names, 'nicknames that they either choose for themselves or are given to them by another thru-hiker based on a personality characteristic or an event' (Rush, 2002: 49). Traditionally, it is common for hikers to be gifted trail names by others; more recently, hikers adopt their own. Sometimes, as with Leslie Rush's choice of 'Turtle' to represent her persistence in hiking the trail (Rush, 2002: 49), this is to emphasize a quality they would like to enact in themselves (or to avoid a moniker with negative connotations). Increasingly, first-time AT hikers' contact with the digital world of hiking representations before they set off leads to pre-emptive selection of trail names. All online journals hosted at *www.trailjournals.com*, for example, prompt their authors for a trail name which is displayed more prominently than their actual name on the site's list of journals and on the pages of their own entries. It is likely that the prominent spaces for hikers' trail names in online journals are less concerned with making them pick a name and more with advertising a name they have already chosen.

An encounter in Miller's hike illustrates a second difference between the affordance of trail registers and online journals, in relation to trail names. Reading in the shelter registers as he hikes, Miller notes that: 'Hikers who use prominent signatures or tidbits of artwork to identify themselves stand out the most … "Wolverine" scratches his signature with multiple strokes, in all caps. The name, and the way it is signed, summons an image of a rugged, macho character, like his namesake in the X-Men movies' (Miller, 2010: 9).

The punchline to Miller's account of Wolverine is that the hiker himself is not a 'rugged macho character' but instead 'a young man … his hiking boots look[ing] too big for his spindly legs' (Miller, 2010: 11). Wolverine's name comes not from his macho nature but from the fact that he is 'wearing the hat of the Michigan *Wolverines*' (Miller, 2010: 9). As geographer David Crouch has argued: 'the repetitive doing of things is affirmative of, and can impel, a powerful sense of being' (quoted in Merriman et al, 2008: 195). In this case, Wolverine repeatedly signing his name impelled on other hikers a powerful sense of who he must be.

In contrasting fashion, journal entries hosted on *www.trailjournals.com* by Greg Claypool, using the trail name Zombie, show a ritualized form of writing and repetition which appears aimed at himself as well as, if not more so than, others. Zombie's online journal entries begin on Monday 14 December 2020 and end on Thursday 21 October 2021. There are breaks in-between, as Zombie is a section hiker rather than a thru-hiker, walking different parts of the Appalachian Trail in different years. The entry quoted next, in full, is typical of this journal:

> Rainy start day [beginning a 'pre-hike']. Planned to go a little farther but it rained all day. Met Happy [another hiker] at shelter. He headed out to continue hiking. Once in dry clothes got in tent and stayed warm. A day in the woods makes me happy! (Claypool, 2021)

These few sentences are filled with evidence of the affective power of writing to shape hikers' practices and promote a ritualized relationship to writing. Zombie writes about being cold and wet at the end of a day hiking, yet still makes the time to post a short journal entry. Particularly notable is the last sentence – 'A day in the woods makes me happy!' – which appears at the end of each of his 52 entries. Following David Crouch and repetition as the source of a sense of self, it is possible that for Zombie, the repetition of writing this same phrase serves both to remind him of the positives inherent in the act of walking, and to become the happy person his blog says he is.

Deliberate connections between wilderness and world

The chief promise of digital detox is not simply that it will allow us a break from our devices and the social media they contain. It is that a detox, or its more quotidian cousin the digital sabbath, can help return to us a sense of control and deliberate, conscious engagement with the digital. As Catherine Price argues in *How to Break Up with Your Phone* (2018: 81), 'Our goal isn't abstinence; its consciousness'. Similarly, Jenny Odell, writing about the need to escape from the attention economy engendered by social media and the digital devices which carry them, suggests the deliberate photographic practice of artist David Hockney, who sought to use technology in new ways to emphasize and so change the lens it offers, as an example to be followed (Odell, 2019: 226). In this final section I want to emphasize the ways in which the Appalachian Trail in the twenty-first century can be understood as a space of digilogue and post-digital practices, a space in-between the analogue idyll and the digital busy-ness of everyday life, by looking at the ways in which a new generation of hikers – fat hikers, hikers of colour, Native American hikers – are using digital devices, technologies, and media on the Appalachian Trail deliberately to change the ways in which the trail is represented in the popular imagination.

Author and activist Rahawa Haile, whom we encountered in our wanderings earlier, has written about the power of social media posts to forge 'an incredible sense of self' on the Appalachian Trail (Haile, 2017a; see also McLaughlin, 2023). For Haile, Instagram provided a means during her Appalachian Trail thru-hike of advertising her efforts to bring books by black authors with her into the wilderness, to reclaim this historically charged landscape for black people. It also provided a platform for her to enact this powerful sense of self through highly public interrelations with her black authors, which spanned both actual and digital spaces. She shows the world to her books, as it were, in emotionally charged events: 'I got to a view. I held them to the light. I told them, firmly, "This is yours"' (Haile, 2017a). By representing these events through Instagram, Haile uses this digital platform to *at once* enact and to express her right to be in Appalachia, as a black woman, on a much larger and arguably more powerful stage.

Haile's mission to use digital devices as a means of putting Appalachia's colonial past front-and-centre through digital devices reflects Lamar Marshall, cultural heritage director of the conservation group Wild South's, long-running Cherokee Roads project (see Marshall, 2016; Schroeder, 2022). This project aims to map the former network of roads and paths through Appalachia used by Cherokee peoples before their forced displacement west. Like the other hikers discussed in this chapter, Lamar uses a rather digilogue mix of 'a GPS device, a few maps, a black notebook, a pen, and firestarter for emergencies' (Moor, 2016b: 163), alongside digital photographs and GIS software to locate, trace and map these ancient trails. Like other authors of colour, Marshall makes purposeful use of online blogs as a means of publicizing his project and attempting to reshape non-indigenous America's understanding of its wilderness areas. One such author is Latria Graham, who has written on the lived reality of people of colour in American rural areas for widely read hiking magazine, *Outside Online*. Graham's punchy title, 'We're here. You just don't see us', appears squarely aimed at white Americans who may buy into the conceit that, as she writes it, 'the black people dislike the outdoors' (Graham, 2018).

This deliberate use of online writings and social media posts to claim a space in the American wilderness in general, and the Appalachian Trail in particular, is not solely the preserve of academics and authors like Haile, Graham, and Marshall. This is apparent in the popularity of an account on Instagram called @UnlikelyHikers, which at the time of writing has almost 2000 posts and 157,000 followers. The Unlikely Hikers are a community of people defined by physical, mental, and social attributes – being fat, being queer, being a parent, being a person of colour – which sets them apart from the highly gendered, highly radicalized world that American hiking still maintains (see Haile, 2017b; Stanley, 2020). Unlikely Hikers is also an Instagram account through which these hikers advertise their presence on the Appalachian Trail and, increasingly, other long-distance trails in America and beyond. In doing so, writes one organizer of this Instagram account, these 'unlikely hikers' are deliberately enacting new ways of being in the hiking world:

> If you're a person of color [sic], or fat, or transgender … you can't just leave that at home … . Hikers who exist within marginalized groups and create Instagram accounts and Facebook pages … to create community and support for each other endure near constant harassment from more privileged hikers who would rather not be inconvenienced by having to admit that these issues are real … . There is no 'staying neutral', in these times. You're either rolling up your sleeves to help or you're part of the problem. (Stanley, 2020: 242)

For Haile and for the Unlikely Hikers, regular posting to social media platforms and writing for online audiences is an act which stands apart from the usual run of smartphone attachment, as digital detoxers might describe it. Catherine Price, for example, sums up the problem of phone and social media use in stark terms:

> We feel busy but ineffective. Connected but lonely. The same technology that gives us freedom can also act like a leash – and the more tethered we become, the more it raises the question of who's actually in control. The result is a paralysing tension: we love our phones, but we often hate the way they make us feel. (Price, 2018: 2)

Despite this general malaise that we may all feel in relation to our digital devices, these hikers of colour, fat hikers, queer hikers, parents, old hikers, and any other hikers one might not expect to find in an American wilderness (where even hiking as a women brings many obstacles [see Heddon and Turner, 2012; Weatherby and Vidon, 2018]), social media is not an unthinking reflex. Instead, it is a powerful, deliberate, *post-digital* tool to effect some form of social change.

Conclusion

The Appalachian Trail has been lauded as a place where people might go to escape from the pressure of the digital attention economy in our everyday lives. As a wilderness site and a place for a long-distance walk, it sits within two rich cultures of escape from the everyday – cultures which have continued to grow in relevance as the 'everyday' increasingly refers to 'the digital'. However, as this chapter has discussed, increasing numbers of hikers are using digital devices on the Appalachian Trail – and, in fact, have been doing so now for at least 20 years. This suggests, as I have argued here, that the trail is not entirely an 'analogue idyll'.

In fact, I argue that the Appalachian Trail is an in-between space, not quite of the everyday world but not entirely divorced from it. Here a digilogue material culture is prevalent, as hikers freely switch from producing and consuming media and representations across a variety of analogue and digital technologies. It is also a post-digital space where 'the digital' as a marker of techno futurist progress is losing its allure, if only for a time, in part due to the practicalities of life in the woods (trees work poorly as charging points, for example). Perhaps there is an argument to be made here for hikers practicing a form of digital detox by accident; certainly, it is a place where hikers can be free to be more deliberate about how they use social media and digital devices.

Two questions emerge from this research. The first relates to the material legacy of the various analogue and digital media hikers produce. Online accounts are easily accessible, but contemporary analogue texts, in the form of shelter registers or journals, remain far less so. Once a spiral-bound notebook is 'used up', they are collected and replaced by local clubs who maintain the Trail and stored away, if not destroyed. This makes it harder than it might otherwise be to compare how things are changing and how the relationship between representations both analogue and digital is developing. The comparisons I offer here are, of necessity, 'before' and 'after' the flood, as it were. Future research requires that these resources are digitized. This is a mission on which I am currently working, in partnership with Jess Cory of the University of North Carolina to digitize these registers (and future registers) and so make them accessible to researchers.

A second question may strike more resonance with geographers, as it concerns the spatial complexity of the twenty-first-century Appalachian Trail as a post-digital place: to what extent do hikers' abilities to share digital representations, especially first-person-view photographs, affect their spatial understanding and practice of the trail? What happens when what hikers see is easily broadcast to all? How can we understand digital representations in terms of the space of the AT, where it begins and ends, and which relations contribute to its co-production? These are questions that cannot be answered here, but which demand continued attention by geographers, and others, to the Appalachian Trail as a digilogue site made of spaces perhaps more complex than the tangled thicket of woods may at first suggest.

References

Abram, D. (2011) *Becoming Animal: An Earthly Cosmology*. Vintage.

Barry, E. (2014) 'Digilog Culture and the Vinyl Revival of the Early 21st Century', *American History Now: Explorations in Digital Curation*, [online] 30 April, Available from: https://americanhistorynow.org/2014/04/30/digilog-culture-and-the-vinyl-revival-of-the-early-21st-century/ [Accessed 15 May 2023].

Brown, T. (2019) 'Harper', in J. Cory (ed) *Mountains Piled Upon Mountains: Appalachian Nature Writing in the Anthropocene*. West Virginia University Press, pp 41–45.

Bryson, B. (1998) *A Walk in the Woods: Rediscovering America on the Appalachian Trail*. Random House.

Calvino, I. (1981) *If on a Winter's Night a Traveler*. Harcourt.

Claypool, G. (2021) 'Journal', *Trail Journals*, [online] Available from: https://www.trailjournals.com/booat2021 [Accessed 15 May 2022].

Cory, J. (2015) 'From Gatewood to Davis: How Technology Changes the Way We Write About the Appalachian Trail', Conference presentation. *Humanities Education Research Association*, San Francisco, CA.

Cramer, F. (2015) 'What Is "Post-digital?"', in D.M. Berry and B. Dieter (eds) *Post-Digital Aesthetics: Art, Computation, and Design*. Palgrave Macmillan, pp 12–26.

Cronon, W. (1996) 'The Trouble with Wilderness: Or, Getting Back to the Wrong Nature', *Environmental History*, 1(1): 7–28.

Digital Detox (2023) 'The Camp Grounded Experience from Digital Detox', *Digital Detox*, [online] Available from: https://www.digitaldetox.com/experiences/camp-grounded/ [Accessed 15 May 2023].

Fondren, K.M. (2016) *Walking on the Wild Side: Long-Distance Hiking on the Appalachian Trail*. Rutgers University Press.

Fondren, K., and Brinkman, R. (2019) 'A Comparison of Hiking Communities on the Pacific Crest and Appalachian Trails', *Leisure Sciences*, 44(4): 403–420.

Frost, L. (2022) 'Family Digital Detox: 5 Truths We Learned on the Appalachian Trail', *Defend Young Minds*, [online] 15 November, Available from: https://www.defendyoungminds.com/post/family-digital-detox-5-truths-we-learned-on-appalachian-trail [Accessed 15 May 2023].

Glebe, R. (2025) 'What Is a Digital Detox – and Why You Need an Unplugged Adventure ASAP', *Adventures in Good Company*, [online] 30 March, Available from: https://www.adventuresingoodcompany.com/what-is-a-digital-detox-and-why-you-need-an-unplugged-adventure-asap [Accessed 14 July 2025].

Graham, L. (2018) 'We're Here. You Just Don't See Us', *Outside Online*, [online] 1 May, Available from: https://www.outsideonline.com/culture/opinion/were-here-you-just-dont-see-us/ [Accessed 15 August 2020].

Gros, F. (2015) *A Philosophy of Walking*. Verso.

Haile, R. (2017a) 'How Black Books Lit My Way Along the Appalachian Trail', *Buzzfeed Reader*, [online] 2 February, Available from: https://www.buzzfeednews.com/article/rahawahaile/how-black-books-lit-my-way-along-the-appalachian-trail#.xx1anBwWZy [Accessed 1 November 2017].

Haile, R. (2017b) 'Going It Alone', *Outside Online*, [online] 11 April, Available from: https://www.outsideonline.com/2170266/solo-hiking-appalachian-trail-queer-black-woman [Accessed 15 January 2019].

Heddon, D. and Turner, C. (2012) 'Walking Women: Shifting the Tales and Scales of Mobility', *Contemporary Theatre Review*, 22(2): 224–236.

Ingold, T. (2006) 'Up, Across and Along', *Place and Location: Studies in Environmental Aesthetics and Semiotics*, 5: 21–36.

Ingold, T. (2010) 'Footprints Through the Weather-world: Walking, Breathing, Knowing', *Journal of the Royal Anthropological Institute*, S121–S139.

Amerson, K., Rose, J., Lepp, A., and Dustin, D. (2020) 'Time on the Trail, Smartphone Use, and Place Attachment Among Pacific Crest Trail Thru-hikers', *Journal of Leisure Research*, 51(3): 308–324.

Maddrell, A., and Della Dora, V. (2013) 'Crossing Surfaces in Search of the Holy: Landscape and Liminality in Contemporary Embodied Practices', *Social and Cultural Geography*, 17(2): 300–321.

Marshall, L. (2016) 'Appalachian Indian Trails', *Wild South*, [online] Available from: https://wildsouth.org/appalachian-indian-trails-by-lamar-marshall/ [Accessed 15 November 2017].

Marx, K. (2019) 'Transgressive Little Pests: Hiker Descriptions of "Shelter Mice" on the Appalachian Trail', *Anthrozoös*, 32(1): 103–115.

McCadams, A. (2022) 'Digital Detox in Asheville: How Switching Off My Device Helped Me Reconnect', *Lonely Planet*, [online] 18 January, Available from: https://www.lonelyplanet.com/articles/digital-detox-in-asheville [Accessed 15 May 2023].

McLaughlin, D. (2023) 'Digital and Non-digital Representations as Actors in the Enactment of Selfhood and Community on the Appalachian Trail', *Social & Cultural Geography*, 24(6): 912–929.

McLaughlin, D. (2022) 'Appalachian Hikers' Digital Journals: Collective Writing for an Unruly Landscape', *Transfers*, 12(1): 84–94.

Mele, C. (2017) 'Levi Felix, a Proponent of Disconnecting from Technology, Dies at 32', *The New York Times*, [online] 12 January, Available from: https://www.nytimes.com/2017/01/12/us/obituary-levi-felix-digital-detox.html [Accessed 15 May 2023].

Merriman, P., Revill, G., Cresswell, T., Lorimer, H., Matless, D., Rose, G., and Wylie, J. (2008) 'Landscape, Mobility, Practice', *Social & Cultural Geography*, 9(2): 191–212.

Miller, D. (2010) *AWOL on the Appalachian Trail*. AmazonEncore.

Minteer, B. (2001) 'Wilderness and the Wise Province: Benton MacKaye's Pragmatic Vision', *Philosophy & Geography*, 4(2): 185–202.

Moor, R. (2016a) 'Here's What Happens to Your Body When You Hike the Appalachian Trail', *Buzzfeed News*, [online] 2 August, Available from: https://www.buzzfeednews.com/article/robertmoor/heres-what-happens-to-your-body-when-you-hike-the-appalachia [Accessed 15 March 2018].

Moor, R. (2016b) *On Trails: An Exploration*. Simon and Schuster.

Muir, J. (1992) *John Muir: The Eight Wilderness Discovery Books*. Diadem.

Nash, R. (2014) *Wilderness and the American Mind*. Yale University Press.

Newton, C. (2018) '"Time Well Spent" Is Shaping Up to Be Tech's Next Big Debate', *The Verge*, [online] 18 January, Available from: https://www.theverge.com/2018/1/17/16903844/time-well-spent-facebook-tristan-harris-mark-zuckerberg [Accessed 15 May 2023].

Odell, J. (2019) *How to Do Nothing: Resisting the Attention Economy*. Melville House Publishing.

Price, C. (2018) *How to Break Up with Your Phone*. Trapeze.

Rush, L. (2002) 'Multiliteracies and Design: Multimodality in the Appalachian Trail Thru-Hiking Community', PhD Dissertation, University of Georgia.

Schroeder, S. (2022) 'Trekking Through Time: Documenting Ancient Cherokee Trails in Western North Carolina', *Media Hub*, [online] 27 April, Available from: https://mediahub.unc.edu/trekking-through-time-documenting-ancient-cherokee-trails-in-western-north-carolina/ [Accessed 20 June 2023].

Scriven, R. (2021) 'A "New" Walking Pilgrimage: Performance and Meaning on the North Wales Pilgrim's Way', *Landscape Research*, 46(1): 64–76.

Shaffer, E. (1983 [1948]) *Walking with Spring*. Appalachian Trail Conference.

Sheffer, R. (2004) 'The Mental Geography of Appalachian Trail Hikers', in K. Harmon (ed) *You Are Here: Personal Geographies and Other Maps of the Imagination*. Princeton Architectural Press, pp 66–69.

Solnit, R. (2002) *Wanderlust: A History of Walking*. Verso.

Stanley, P. (2020) 'Unlikely Hikers? Activism, Instagram, and the Queer Mobilities of Fat Hikers, Women Hiking Alone, and Hikers of Colour', *Mobilities*, 15(2): 241–256.

Stelter, T. and McCrickard, D. (2017) 'Hiking the Appalachian Trail with Technology', *NatureCHI 2017*, workshop at MobileHCI017, 4 September 2017, Vienna, Austria, 1–5.

Swift, E. (2015) 'Murder on the Appalachian Trail', *Outside Online*, [online] 2 September, Available from: https://www.outsideonline.com/2011326/murder-appalachian-trail [Accessed 15 November 2017].

Syvertsen, T. (2022) 'Offline Tourism: Digital and Screen Ambivalence in Norwegian Mountain Huts with No Internet Access', *Scandinavian Journal of Hospitality and Tourism*, 22(3): 195–209.

Tesson, S. (2022) *On the Wandering Paths*. University of Minnesota Press.

The A.T. Guide (2023) *The A.T. Guide: A Handbook for Hiking the Appalachian Trail*, [online] Available from: https://www.theatguide.com/ [Accessed 15 November 2017].

Thompson, N. (2020) 'A Nameless Hiker and the Case the Internet Can't Crack', *WIRED*, [online] 2 November, Available from: https://www.wired.com/story/nameless-hiker-mostly-harmless-internet-mystery/ [Accessed 15 May 2023].

Weatherby, T. and Vidon, E. (2018) 'Delegitimizing Wilderness as the Man Cave: The Role of Social Media in Female Wilderness Empowerment', *Tourist Studies*, 18(3): 332–352.

Wylie, J. (2016) 'A Landscape Cannot Be a Homeland', *Landscape Research*, 41(4): 408–416.

5

Finding Digital Distance: Digital Detox and Conspicuous Disengagement

Paul O'Connor

Introduction

This chapter proposes the notion of digital distance to capture the ways in which people seek to manage their use of digital media devices. It highlights both the valorization of voluntary digital disconnection while underlining the impracticality of disengagement for many individuals simply wishing to pursue some control over the way they use their devices such as smartphones and laptops. The increasing frustration with digital life is neatly identified in Zeena Feldman's (2021) discussion of a 'beyond choice ontology' where most individuals feel their daily life is permeated with limited agency to disconnect. This lack of choice prefigures a notion of privilege as part and parcel of the process of disconnection and signals a new conceptual distinction in the digital divide. It highlights that, increasingly, those who still have a choice, those most able to disconnect are also those with forms of wealth and privilege that enable them to circumvent the more quotidian traps of digital dependence. They inhabit a different reality to those in the 'beyond choice ontology'. In concert with these developments a parallel issue of laudable distinction has emerged associated with the need to exercise control over digital devices and care for one's own wellbeing, productivity, and time management. Here a clear irony exists in the fact that these devices exist as tools to aid personal organization and effective use of time. This laudable disconnection is framed as conspicuous disengagement, an increasing interest in the esteem associated with limiting, or limited use of digital media devices. Sociologically, it notes the development of a new form of distinction and meritocracy surrounding the notion of voluntary digital disengagement.

It argues that digital detoxes have become an esteemed token of wellbeing and mental health and in turn signal a new digital divide, not between the haves and have nots of technology, but between those with freedom to disconnect and those who feel that they cannot disconnect, and perhaps more urgently, that they feel they should disconnect.

A new digital divide

For more than two decades educators and policy makers have pushed for digital inclusion and literacy, promoting being connected as a necessary tool for survival in the twenty-first century (Compaine, 2001; O'Hara and Stevens, 2006; Segev, 2010). There has been increasing critique of the digital divide as too simplistic a term, flawed in capturing the vast differences in power, wealth, and cultural capital of users (Willis and Tranter, 2016). Indeed, the normative frame of the digital divide proposes that everyone needs and requires technological connection. This has manifested in an increasingly digital public sphere where navigating public services such as public transport, health care, education, and access to urban infrastructure is mediated by smartphones. A further critique of the way 'closing the gap' in the digital divide has created a path dependence of digital overload relates to the fact that there has long been a lack of diverse research on the topic. Van Dijk (2006) argues that the divide has been framed as a static issue and overlooked the complexity of those who are not 'have nots' but 'want nots'. Additionally, he argues that the digital divide is shaped largely by quantitative and policy-framed research and there is a paucity of qualitative work identifying nuances in how the divide manifests in people's everyday lives. A key complaint the authors offer is that there needs to be more conceptualization within the divide.

In response to these criticisms, this chapter presents both a qualitative response to the digital divide and develops the conceptual frame of 'digital distance' that speaks directly to the already embedded notion of space in digital and technological engagement via the frame of the 'divide'. Digital distance responds to what Ragnedda (2018) has suggested as the third digital divide, the interaction between social and digital capital. Early explorations of the digital divide focussed on access to digital and online technologies. Material inequalities clearly conformed with poor access to digital technologies. As the story has progressed, the second digital divide has addressed the uses of technology, how they are harnessed and manipulated for knowledge, wealth, and social engagement. What is now increasingly apparent is that forms of prestige are being linked to time spent offline. As access and competencies have increased, so too has distaste for certain aspects of digital engagement that appear passé, gauche, or a threat to emotional and physical wellbeing. Sociological notions of social capital and conspicuous consumption dovetail with self-employed strategies to create digital distance. The most overt and visible form of disconnection comes from the health and wellbeing oriented 'digital detoxes'.

The digital detox is an emerging activity that exists in opposition to much of the rhetoric surrounding digital inclusion. The constant push for more connectivity is regarded by some people as short-sighted and out of step with the practical needs of individuals, families, and the workings of institutions such as schools and offices (Gregg, 2011). This can also be linked to a push in academic work that has become focal on *big data* and *smart cities* (Obar, 2015; Picon, 2015; Custers, 2016; Pileggi, 2016; Barns, 2017; Lake, 2017; Pettit et al, 2017; Vu and Hartley, 2017). A more acute focus on the overloading of digital learning and online living became a hotly debated topic during the lockdowns of the COVID-19 pandemic in 2020/21. The vernacular of digital life during the early stages of the pandemic included cognitive overload, doomscrolling, and Zoom fatigue, all synonymous with too much time in digital spaces (Williamson, Eynon, and Potter, 2020; Schmitt, Breuer, and Wulf, 2021; Aagaard, 2022).

The concept of digital distance recognizes that digital technology users face a variety of contradictions and paradoxes. Devices have come to limit our physical gestures and the variety of materials that populate our daily lives (Nova et al, 2012). People work, socialize, shop, and relax in front of the same devices. Accordingly, the notion that we should limit or distance our time from devices may be a prosaic fact. It is however complicated by the fluid way in which these devices occupy our daily lives, and in many situations come to be essential. For someone who is insulin dependent their phone is more than a necessity, it is a life support system monitoring their blood glucose. The Dexcom app inevitably sits alongside other less critical services on the user's smartphone, amid email, TikTok, Spotify, and Apple Pay. Thus, the critical and the superficial are encased in the same space, with no physical distance or material architecture separating them. Just as wearable technology aids people in keeping fit and monitoring their medical needs, Apple's screen time monitoring that was introduced in 2018 signals that users of networked digital technologies want more control over their devices and data in line with the quantified-self movement (Lupton, 2016). Yet these efforts directly challenge the attention economy on which brands, social media, and apps depend (Zuboff, 2019). The move towards digital distance recognizes a compromise between being connected and being overwhelmed and highlights that strategies focussed on how to manage time online differ enormously between individuals.

Recent research has noted a growth in the numbers of people using social media to indicate the start and end of voluntary periods of time offline with hashtags and geotagged photos (Jorge, 2019). In the last few years a variety of popular literature has emerged that focusses on the need to disengage from digital life, to nurture yourself and your children beyond the glare of screens. Many of these publications can be regarded as self-help books with pithy titles like *How to Break Up with Your Phone* (Price, 2018) and *Off: Your Digital Detox for a Better Life* (Goodin, 2018). These, and numerous other titles (Fielding-Banks, 2015; Aiken, 2017; Ellis, 2017) are advocating the benefits of digital detoxes. A digital detox industry has emerged with spas, nightclubs, retreats, and lifestyle merchandise branded to promote these conspicuous forms of disengagement (Fish, 2017; Karppi, 2018). Digital disengagement has quickly been adopted as a playful and ironic motif to market merchandise (Jorge, 2019). Consumption of T-shirts and mugs with slogans such as 'offline', or 'JOMO' (Joy of Missing Out) are potent examples of conspicuous disengagement. Merchandise and social media posts announcing digital detoxes highlight that it is not simply good enough to disengage from social media, people also need to know you disengaged. The 'announcement' dovetails with performative practices of social media, as a confessional space of display about the self, self-care, and wellbeing. While digital detoxes may be held in high esteem, for many people they are impractical and unobtainable in the long term. Research in this field has repeatedly identified that voluntary breaks from social media can be energizing but not life changing (Fish, 2017; Karppi, 2018; Jorge, 2019). It appears that only those who have been able to impose a complete disconnection with online media devices, in effect no online presence, experience a transformation through disconnection. However, as research has identified (Hardey and Atkinson, 2018), such voluntary disconnectors occupy a privileged position and accordingly experience considerable obstacles in maintaining such a lifestyle. The challenges of disconnection indicate that the parameters of the digital divide have expanded (Willis and Tranter, 2016); now in the global north, disconnection may indeed be aspirational. It is also unachievable to growing numbers of people who feel the burden and necessity in being online and accessible 24/7.

Method

This research began as a pilot study with students in my sociology of media and digital society classes. Over the course of two years, I ran two digital detox experiments with students and collected their feedback and responses. I draw on these early insights in my analysis. The project was extended to include qualitative interviews with participants who had individually chosen to perform digital detoxes. First, I recruited participants through an online questionnaire that I distributed on Twitter, Tumblr, LinkedIn, and Facebook. I also printed flyers which I distributed at two local universities in Hong Kong and also at one Sociology conference in Toronto between July 2018 and April 2019. The survey was used to gather some basic demographics about digital detox, and to convert respondents into participants in qualitative interviews. The survey yielded 17 responses: 11 female, and six male. Eleven were from Europe, four from Hong Kong, and two from South America. The age distribution was broad with four aged 18–25, seven aged 25–35, and six aged 35–50. I was able to recruit 14 participants for interviews, five directly through the surveys and a further nine through snowball recruitment and word of mouth surrounding my research project. The majority of my interviews were with British participants and were conducted either in person or via Skype. The small sample highlights the challenge in identifying a broad collection of individuals concerned about the time they spend online, and in turn asking for online responses on the topic, and interview time. I took notes from these interviews and wrote up my findings following our discussions; in one case following up with some supplementary emails. The coding and analysis of the surveys and qualitative interviews were interpreted alongside tracking of social media posts on digital detox and self-help books on digital disengagement. The research was approved by Lingnan University ethics board.

The complexity of researching digital disengagement resulted in the mix-method design of the project. The most important focus was to recruit individuals who were negotiating between complete disconnection as is observed in Hardey and Atkinson's (2018) research, and forms of limiting use. Due to the nature of the topic, it was also a key concern to recruit beyond online platforms, though this proved limited. Notably, on the occasions in which I presented my research I was consistently able to recruit additional participants through word of mouth and snowball participation. The research was informed by a working hypothesis that proposed that digital detoxes were difficult for most people to achieve.

The evolution of the digital detox

The short history of the mobile phone has seen a meteoric rise in its popularity and ubiquity. With the introduction of the iPhone in 2007, the smartphone has become a necessity for the majority of denizens in developed countries across the globe. At the end of 2017 it was estimated that there were two billion smartphone users globally (Statista, 2018a). As of 2018 the percentage of the population using smartphones is 69.5 per cent in France, 73 per cent in the United Kingdom, 77 per cent in the United States, and 79 per cent in Hong Kong (Pew Research Center, 2018; Statista, 2018b). In concert with this growth in popularity, there has been an accompanying fetish tied to ever newer models of smartphones, their ever-improving functions, and their role as status symbols. Until recent years a media frenzy accompanied the release of each new iPhone model with news reports focussing on eager consumers sleeping outside of stores in anticipation of the latest model. Some have come to associate a religious-like devotion to the handsets (Campbell and La Pastina, 2010; Robinson, 2013). Smartphones have become 'must have' items for many and the latest models, along with customizations, are imbued with social significance. Research has identified the conspicuous consumption of smartphones amongst Nigerian college students (Liadi, 2016), Chinese migrant workers (Huang and Wang, 2018), and also indicated that men become less concerned with having the most recent iPhone once they enter into long term relationships (Hennighausen and Schwab, 2014). Even smartphone apps, like *I am Rich*, with expensive download prices have been developed so wealthy iPhone users can distinguish themselves among the less affluent (Campbell, 2015).

Smartphones are not a vanity item alone, as an exemplar of the notion of technological and media convergence the smartphone has displaced countless objects facilitating mobility (Urry, 2007; Elliott and Urry, 2010) in fluid and new ways (Greenfield, 2017). Smartphones have also become a central way in which we measure and quantify our social bodies, a tool to measure health, bodily aesthetics, and physical change (Lupton, 2016). While there has been lengthy discussion surrounding the health and social risks of smartphone use, the growth in smartphone popularity has continued unabated and has transformed the way individuals behave and how cities, their transportation infrastructure and businesses, operate (Greenfield, 2017; Graham, 2018). Yet, despite the popularity and sophistication of these devices, a new era of self-enforced disengagement from them has emerged.

The digital detox has become a phrase readily associated with a therapeutic break from digital life. It is most simply described as a period of time away from networked digital devices. These are typically understood to be smartphones, and to a lesser extent personal computers, tablets, or all forms of screens and digitally networked devices. The need to pursue a digital detox is inevitably framed as an issue of subjective wellbeing, a need to be free from the constant demands of work, to achieve some self-esteem beyond social media, to engage with real people in the real world. Accompanying the notion of a digital detox are a host of implicit values, these being that smartphones are addictive, difficult to live without, ultimately unrewarding, and a potential threat to wellbeing. The process implies that our digital society is in some ways toxic, and pathologizes our uses of digital technology. The language here relates to how we consume social media, that it is in some way inauthentic, akin to processed and junk food. The toxicity of digital life abounds with food metaphors (Sutton, 2017), diets, and nutrition (Brewer, 2022). Here a connection can be made to the sociology and anthropology of consumption. As Douglas and Isherwood note (1979), acts of material consumption and marketing have garnered a depth of value judgement that is somewhat confused. Goods that we consume, they argue, serve a social and cultural importance as a field of communication, they are cultural acts played out in material artefacts. Combining Douglas and Isherwood's symbolic understanding of consumption with Sutton's food metaphors, it is evident that the consumption of social media, or digital life, is neither good nor bad, but another aspect of our complex, symbolic, and layered cultural communications and expressions.

The phrase 'digital detox' was added to the Oxford English Dictionary in 2013 (Kennedy, 2013), notably the same year FOMO (Fear of Missing Out) was first included. However, the use of the phrase can be found online as early as 2003. The earliest discussions on digital detox are focussed on mental health and internet addiction, aiding users to find a way out of compulsive uses. In 2008, Adbusters promoted digital detox as part of their long running 'turn off TV week'. Kalle Lasn, the founder of Adbusters, argued about the use of the term at this early stage, that a digital detox was a healthy activity that could provide people with insight to their level of technological addiction (Cisar, 2009). Google Trends (2020) shows that 'digital detox' grew notably in popularity as a search term in 2012 and its popularity has continued to grow globally ever since.

There is a growing collection of research on forms of voluntary digital disconnection, or digital detox (Barchilon Ben-Av and Ben-Av, 2016; Fish, 2017; Karppi, 2018; Kushlev and Dunn, 2018; Jorge, 2019). Research has highlighted that motivations to be online are contingent on a host of important connections. Some individuals might surely be experiencing the condition of FOMO (Dossey, 2014), but others are truly compelled to be online for their jobs, careers, relationships, and health. Some research that has looked at digital detox has proposed it as a solution for deteriorating learning environments with students phubbing (a portmanteau of both phone and snubbing) teachers (Ugur and Koc, 2015), and the necessity of taking seriously those who wish to be unconnected (Silk et al, 2016).

One study of prolonged disconnection by Hardey and Atkinson (2018) researched an international cohort of 50 participants over five years who had entirely disconnected from online life. These were individuals who in many cases did not possess a mobile phone, did not have any social media presence, and privileged face-to-face contact and local community. The researchers recruited participants by placing advertisements in the *London Review of Books*. Employing a novel methodology for the twenty-first century, some participants corresponded by post in order to contribute to the study. Implicit in the study is the nascent notion of digital distance, echoing in the methodology of handwritten communiques, and in the distance at which participants were able to place themselves from technology. The distance participants had in their study can also be read as a social distance that reveals them as relatively privileged individuals able to circumvent particular modes of work and online communication. The research highlights that those who are disconnected experience a host of difficulties in performing seemingly simple activities such as buying a train ticket. They note that without being connected, without possessing an email, they become an anomaly. They are marked by distinction, they are othered from the ubiquity of digital life. The motivations for such extreme levels of digital disconnection are multiple but pertain to concerns about safety in a post-Snowdon world, privacy, and a desire to be engaged with the *real world*. An important feature of Hardey and Atkinson's work is that it relies on recruiting a fairly well educated and affluent demographic, that is, readers of the *London Review of Books*. As a result, it appears that many of the respondents are wealthy enough to be able to order their lives without the necessity of digital connections. I also argue that the distinction of such a disconnection has become associated with a host of positive disconnections; wealth, leisure, freedom, independence, time and so on. These values are the inverse of those more commonly described in the rhetoric of the digital divide, where disconnection is associated with lack of access, affluence, and autonomy. The participants in Hardey and Atkinson's study appear to not be the same individuals that the tide of new digital detox self-help books seem to be aimed at. It may therefore be argued that disconnection has become, at least in varying measures, somewhat aspirational.

The rise in books guiding consumers to a safe level of disengagement with technology is notable. From the celebrated and cerebral *Alone Together* (Turkle, 2012), to the pithy titled *Irresistible* (Alter, 2018), there are now numerous self-help books on managing digital addiction (Fielding-Banks, 2015; Aiken, 2017; Ellis, 2017; Goodin, 2018; Price, 2018; Zahariades, 2018). In a content analysis of these books, many conform in style and tone to self-help books, providing a framework and schema to apply lifestyle change in overcoming a problem. Bergsma (2008) argues that self-help books are commonly ridiculed for being poorly written and opportunistic. What is remarkable about the proliferation of these digital detox texts is that they speak to a growing number of people who feel like they have lost control over technology in their own lives and those of their family members. While some of these texts provide a guide, others offer informed analysis on the sometimes-pernicious effects of technology and identify that there is not necessarily a quick fix. What all of them share is a recognition of the need to limit our use of and time spent with certain forms of devices, apps, and software. Yet each of these books are available in digital copies accessible via Kindle and Apple Books.

A consistent feature of research on voluntary digital disconnection is the identification of the paradox and irony of digital detoxes. Jorge (2019) provides an in-depth review of the ways in which digital detoxes are performed online. She argues that digital detox posts on social media conform to three tropes, they are concerned with exercising some autonomy over wellbeing, disconnection proselytization, and an emphasis on disconnection as a lifestyle or marketing opportunity. In Jorge's research Instagram posts often embrace the irony of digital detox in a tongue-in-cheek way. In one post a holiday photo is shared with the user wearing a T-shirt with the slogan 'Offline I am on Vacation', and another has a man taking a selfie in bed wearing an eye mask with the word 'Offline' printed on it.

Conspicuous disconnection

One of the most salient issues in discussions on digital detox is the way in which the practice has become esteemed. This is quite remarkable because it highlights a rather quick shift in attitudes to technology and the prestige of being online and connected. With reference to Thorstein Veblen's (2005) influential concept of conspicuous consumption I contrast a number of ways in which digital detox is imagined. This influential paradigm has been used to show how material consumption of goods and brands provide symbolic forms of capital, pertaining to inferred wealth, status, or subcultural niche. A material example of conspicuous non-consumption exists in the adbuster branded 'Black Spot Unswoosher' shoe that embraces recyclable materials, Klein's 'No Logo' ethos, and an anti-Nike, anti-Sweatshop symbolism (Adbusters, 2018). The concept of online conspicuous non-consumption has been proposed by Portwood-Stacer (2013) in reference to the cultivation of taste and distinction by users who shun Facebook. This seminal research identifies both the lifestyle choice of disconnection and the way in which it can be read as a performative and even moral act. In proposing an augmentation to this schema, conspicuous disconnection also addresses changing attitudes to not just the media, but also digital devices.

A remarkable development in the conspicuous uses of smartphones has been the way in which disengagement from them has evolved as an esteemed lifestyle and mental health choice (Portwood-Stacer, 2013; Jorge, 2019). The paradox is that frequently social media is a place where forms of digital detox are promoted and made glamourous. Niomi Smart, a famous British fashion and lifestyle blogger with over 1.3 million Instagram subscribers, and 1.5 million YouTube subscribers, makes her living through social media. She is typical of many of the influencers who have cultivated lucrative brand endorsements and partnerships (Duffy, 2017). In one Instagram post Niomi is photographed in black and white by a pool, topless with her arms obscuring her breasts while she reads a book. In the post she declares that the photo was of her taking time off from the digital world and advocating digital detoxes (niomismart, 2018). The post draws on a number of symbolic elements; she promotes the allure of the digital detox by posing at a luxury resort (geotagged) and displaying her slim and toned body juxtaposed with book and Olympus camera. The photo is evocative of the tropes identified by Duffy and Hund (2015) drawing on the 'glam life', curated sharing about personal life, and also Niomi's passion for blogging.

Smart also acknowledges the paradox of announcing digital detoxes on social media in an editorial for *Glamour Magazine* (Smart, 2019). At the time of writing Instagram lists over 150,000 posts hashtagged #digitaldetox. Browsing through these photos provides an insight to the posed conspicuous nature of the digital detox, people out in nature, spending time with their families, eating great food. Similarly, the comments to these posts are often full of celebrations, sometimes envy, at the ability to disconnect and take time for yourself. I argue that the increasing prestige of disconnection and the way it is marketed online has a consequence of making some individuals feel that they are flawed or incapable of disconnection.

The branding of digital detoxes has developed in concert with their rising popularity. In 2016, Dolmio, a pasta sauce brand, began to produce pepper mills that would turn off electric devices, marketed as a boon for family supper times (Roper, 2016). The Mandarin Oriental hotel chain also promote their digital detox spas where users can indulge in relaxation and pampering without the intrusions of technology (Kufer, 2018). There are also various locations and events that promise 'phone free' socializing. The brand *You Had to Be There* specializes in party events where smartphones are locked away, the rules are simply no phones and no pictures. Their branding is focussed not just on digital detox but also the other spectre of the digital age FOMO. From a conspicuous disconnection perspective the brand provides a problem: how do people know that you have disconnected and are partying? In order to sate the need to promote the disconnection and consume at the same time, the brand markets sweaters and shirts that you can buy online at their website (You Had to Be There, 2019). Such branding highlights that digital detoxes are desirable and enmeshed with forms of conspicuous consumption, and despite their focus on disconnection, are promoted, documented, and celebrated online.

Digital distance

The esteemed nature of the digital detox was one that was shared by my participants. Out of the 14 interviews I performed, ten participants appeared to be keen to share with me their strategies for limiting the amount of time spent online. In many cases these seemed to translate as pride, that they were not susceptible to the pernicious effects of technology. The remaining four were by contrast confessional and were concerned that their use of digital technology was out of control. Implicit in the responses of all of my participants was the association that spending less time online, scrolling through their phone, or looking at screens was in some way laudable. It is indeed a notable feature and flaw of the design of this project that those who were not interested in digital detoxing, did not respond. However, the majority of the people I did speak to were similarly those engaged in conspicuous disconnection as they were happy to discuss their strategies and techniques of digital detox.

In analysis of interviews it became apparent that all 14 of my respondents were alike in the fact that none of them performed a sustained digital detox, spending a day, weekend, or longer without their phone or other devices. All the participants dismissed this as largely impractical and not necessary. However, all of them were engaged in strategies to limit or distance themselves from their devices at particular times of the day. Sophie from the UK, for instance, a recently single lawyer in her fifties, was mostly concerned with spending too much time on Facebook in the evenings (where she would come across her ex), and reading the *Daily Mail* online. Her strategies were novel, she decided to block the URL of the *Daily Mail* website so even if she was tempted to look at it, she could not access it. In order to circumvent spending time on Facebook, she had purchased a projector so that she would watch movies from her laptop and then be unable to access the web browser at the same time. It is notable here that a total digital detox is undesirable. Sophie wants to watch Netflix but not browse the web. Perhaps here the food metaphors of Sutton (2017) contrast, where one does not want to cease eating, but simply be selective about certain foods. Distance is also relevant in the notion of distinction, in that by distancing herself from the gauche 'sidebar of shame' (Brown, 2012) on the *Daily Mail* website, she was also distancing herself from a type of identity and personhood.

Amy, a tech worker in her mid-thirties and also from the UK had a more low-tech approach which can be described quite literally as digital distancing. At home she intentionally only has one phone charger and she leaves it in her bedroom. Her phone is upstairs whenever she is home, and if she needs it, she must climb the stairs to go and use it. Other participants spoke of similar strategies, deleting apps so they would not waste their time or hiding technology in drawers and cupboards so as not to distract them as they went about their daily chores.

Yuki, a 19-year-old student on exchange in Paris had a different expression of digital distance to other participants. Far from home and in a new city she was increasingly frustrated that her day to day was identical to life back home. She simply spent time alone in her room, watching videos on her laptop or scrolling through social media. Her issue of distance was that she felt her world was too small and claustrophobic and she was searching for ways to break the tie she felt to her devices and immerse herself in the new culture and city she was now living in.

For several participants, a desire to limit their time online was focal on limiting their presence online. Social media, Facebook, and LinkedIn were seen to be distractive 'time suckers' but also necessary evils keeping a connection to new opportunities, be they work or social. One respondent in her early twenties had deleted her Facebook after she graduated and entered the world of work feeling that her decade of photos on her profile were no longer representative of who she was. Compounded in these assessments was also an ambivalence about the surveillance economy in which participants were sometimes helplessly cast. All my participants were aware of the Cambridge Analytica scandal of 2016, though none expressed any deep concern about how this affected their everyday life. One participant argued that she intentionally limited what she shared online to circumvent the insidious nature of surveillance.

Two male participants informed me that they had used old Nokia phones as a way to avoid spending too much time online during working hours. In both cases the men had not surrendered smartphones entirely, they just strategically swapped their sim cards into the more low-tech Nokia phones when they needed to avoid the temptation of using their phone. While both of these men were more concerned about the amount of time spent online, one openly admitting that he was addicted, they did not appear to be spending any more time on their devices than other participants. Thus, feeling good about your digital engagement appears to be more important than the time spent on devices. This indicates that the level of disconnection that Hardey and Atkinson (2018) studied is actually unobtainable for many people. The necessity to be connected for work, parenthood, travel, and social life is now simply too ubiquitous for most people to entirely disconnect.

The mundane nature of our digital dependence was made acutely aware to me when I asked students to perform a digital detox, an activity in which I also participated. To my surprise I was unprepared for the amount of stress and disruption that it would cause my students. One 20-year-old male student was unable to catch a bus without his phone, never having learnt the bus numbers or timetable for his destination; another could no longer follow their university timetable without their phone. Other students lost work opportunities, and even had relatives fraught with concern over their wellbeing because of sustained hours of disconnection. What then appears to be more realistic as well as more manageable is not the health-oriented detox, but the more practical distancing in use of digital technologies. Apple's screen time monitoring (Solon, 2018) provides a wholly quantitative assessment of technology use; it is clear that users derive a considerable amount of quality from their devices' facilitation of an array of everyday chores with remarkable efficiency and cost effectiveness. This dependence of technology does however indicate a new finding, that of the emergence of a new form of digital divide.

Conclusion

The concept of digital distance highlights a new expression of the digital divide where the ability to truly disconnect has become a niche pursuit and attainment in our digital society. This is a significant rearticulation, inverting previous understanding of the digital divide where disconnection is a social flaw, an obstruction to education, progress, and work, and now recasting disconnection as aspirational, healthy, and, perhaps most ironically, productive. This research contributes to the sociological debate on how the digital detox has arisen as an esteemed form of conspicuous disconnection from technology. It argues that disconnection for most people is not just impractical but also undesirable. Feldman (2021: 105) frames this as both the necessity and impossibility of disconnection. This impossibility is further confirmed by the fact that the digital detox is most identifiable as an online phenomenon, documented, hashtagged, and 'liked' by millions of users online. In qualitative work with participants it appears that strategies of digital distance strike a balance from an all-out disconnection with networked digital technologies and a negotiated use. As contemporary life is now so deeply entwined with digital technologies, many people who attempt a digital detox may simply want to avoid their phone but still watch Netflix, or be online and accessible for work opportunities, or family members. As an extension of the debate on digital inclusion, the right or need to disconnect will become a pressing issue for those that who wish to address the issue of digital distance.

References

Aagaard, J. (2022) 'On the Dynamics of Zoom Fatigue', *Convergence*, 28(6): 1878–1891.

Adbusters. (2018) 'The Blackspot Unswoosher', *Adbusters*, [online] Available from: https://subscribe.adbusters.org/products/blackspot [Accessed 12 March 2023].

Aiken, M. (2017) *The Cyber Effect*. John Murray.

Alter, A. (2018) *Irresistible: The Rise of Addictive Technology and the Business of Keeping Us Hooked*. Penguin.

Barchilon Ben-Av, M. and Ben-Av, R. (2016) 'Smartphones: From Distraction to Attraction', *Journal of Educational Technology Systems*, 45(1): 93–102.

Barns, S. (2017) 'Smart Cities and Urban Data Platforms: Designing Interfaces for Smart Governance', *City, Culture and Society*, 12: 5–12.

Bergsma, A. (2008) 'Do Self-help Books Help?', *Journal of Happiness Studies*, 9: 341–360.

Brewer, J. (2022) 'Digital Nutrition', Jocylen Brewer, [online] Available from: https://jocelynbrewer.com/digital-nutrition/ [Accessed 2 July 2023].

Brown, A. (2012) 'The Shocking Thing about the Mail Online's Sidebar of Shame', *The Guardian*, [online] 24 March, Available from: https://www.theguardian.com/commentisfree/2012/mar/24/mail-online-sidebar-of-shame [Accessed 12 July 2023].

Campbell, C. (2015) 'Status Matters? Contradictions Surrounding Conspicuous Consumption', in A.N.G. Martínez (ed) *Being Human in Consumer Society*. Ashgate, pp 41–65.

Campbell, H.A., and La Pastina, A.C. (2010) 'How the iPhone Became Divine: New Media, Religion and the Intertextual Circulation of Meaning', *New Media & Society*, 12(7): 1191–1207.

Cisar, K. (2009) 'Adbusters Founder Says "Digital Detox" Can Provide Epiphany', *Madison*, [online] 12 June, Available from: https://madison.com/entertainment/city_life/adbusters-founder-says-digital-detox-can-provide-epiphany/article_5cf1afca-08ea-5b26-ae31-76f400a1db6c.html [Accessed 14 July 2023].

Compaine, B.M. (2001) *The Digital Divide: Facing a Crisis or Creating a Myth?* MIT Press.

Custers, B. (2016) 'Click Here to Consent Forever: Expiry Dates for Informed Consent', *Big Data & Society*, 3(1): 1–6.

Dossey, L. (2014) 'FOMO, Digital Dementia, and Our Dangerous Experiment', *Explore*, 10(2): 69–73.

Douglas, M. and Isherwood, B. (1979) *The World of Goods: Towards and Anthropology of Consumption*. Routledge.

Duffy, B.E. (2017) *(Not) Getting Paid to Do What You Love: Gender, Social Media, and Aspirational Work*. Yale University Press.

Duffy, B.E. and Hund, E. (2015) ' "Having it All" on Social Media: Entrepreneurial Femininity and Self-Branding Among Fashion Bloggers', *Social Media + Society*, 1(2): 1–11.

Elliott, A. and Urry, J. (2010) *Mobile Lives*. Routledge.

Ellis, M. (2017) *Digitox: How to Find a Healthy Balance for Your Family's Digital Diet*. Culture Transform.

Feldman, Z. (2021) 'Quitting Digital Culture Rethinking Agency in a Beyond-Choice Ontology', in A. Chia, A. Jorge, and T. Karppi (eds) *Reckoning with Social Media: Disconnection in the Age of the Techlash*. Rowman & Littlefield, pp 103–130.

Fielding-Banks, O. (2015) *Unplugged: The Essential Digital Detox Plan*. Carlton Books.

Fish, A. (2017) 'Technology Retreats and the Politics of Social Media', *tripleC: Communication, Capitalism & Critique*, 15(1): 355–369.

Goodin, T. (2018) *Off: Your Digital Detox for a Better Life*. Octopus Publishing Group.

Google Trends. (2020) 'Digital Detox', *trends.google.com*, [online] Available from: https://trends.google.com/trends/explore?date=all&q=%22digital%20detox%22 [Accessed 14 April 2023].

Graham, S. (2018) *Vertical: The City from Satellites to Bunkers*. Verso.

Greenfield, A. (2017) *Radical Technologies*. Verso.

Gregg, M. (2011) *Work's Intimacy*. Polity.

Hardey, M. and Atkinson, R. (2018) 'Disconnected: Non-Users of Information Communication Technologies', *Sociological Research Online*, 23(3): 553–571.

Hennighausen, C. and Schwab, F. (2014) 'Relationship Status Moderates Men's Conspicuous Consumption of Smartphones', *Letters on Evolutionary Behavioral Science*, 5(2): 13–16.

Huang, Z. and Wang, C.L. (2018) 'Conspicuous Consumption in Emerging Market: The Case of Chinese Migrant Workers', *Journal of Business Research*, 86: 366–373.

Jorge, A. (2019) 'Social Media, Interrupted: Users Recounting Temporary Disconnection on Instagram', *Social Media + Society*, 5(4): 1–19.

Karppi, T. (2018) *Disconnect: Facebook's Affective Bonds*. University of Minnesota Press.

Kennedy, M. (2013) 'Omnishambles among New Words Added to Oxford Dictionaries Online', *The Guardian*, [online] 28 August, Available from: https://www.theguardian.com/science/2013/aug/28/omnishambles-new-words-oxford-dictionaries [Accessed 14 April 2023].

Kufer, K. (2018) 'The Digital Detox: Mandarin Oriental Digital Wellness Escape', *Harper's Bazaar Arabia*, [online] 6 May, Available from: https://www.harpersbazaararabia.com/lifestyle/the-escape/the-digital-detox-mandarin-oriental-digital-wellness-escape [Accessed 17 March 2023].

Kushlev, K. and Dunn, E.W. (2018) 'Smartphones Distract Parents from Cultivating Feelings of Connection When Spending Time with Their Children', *Journal of Social and Personal Relationships*, 36(6): 1619–1639.

Lake, R.W. (2017) 'Big Data, Urban Governance, and the Ontological Politics of Hyperindividualism', *Big Data & Society*, 4(1): 1–10.

Liadi, O.F. (2016) 'College Students and Smartphone Ownership: Symbolic Meanings and Smartphone Consumption Among Nigerian Students', *Communication*, 10(1): 17–31.

Lupton, D. (2016) *The Quantified Self*. Polity Press.

niomismart (2018) 'Swapping My Phone for a Book', Instagram, [online] 18 October, Available from: https://www.instagram.com/p/BpFTpzSh7EA/ [Accessed 19 March 2023].

Nova, N., Miyake, K., Chiu, W., and Kwon, N. (2012) 'Curious Rituals', *The Near Future Laboratory*, [online] 1 January, Available from: https://nearfuturelaboratory.com/projects/en/curious-rituals/ [Accessed 9 July 2023].

O'Hara, K. and Stevens, D. (2006) *Inequality.com: Power, Poverty and the Digital Divide*. Oneworld.

Obar, J.A. (2015) 'Big Data and The Phantom Public: Walter Lippmann and the Fallacy of Data Privacy Self-management', *Big Data & Society*, 2(2): 1–16.

Pettit, C., Bakelmun, A., Lieske, S.N., Glackin, S., Hargroves, K., Thomson, G., et al (2017) 'Planning Support Systems for Smart Cities', *City, Culture and Society*, 12: 13–24.

Pew Research Center. (2018) 'Mobile Fact Sheet', *Pew Research Center*, [online] 13 November, Available from: http://www.pewinternet.org/fact-sheet/mobile/ [Accessed 18 July 2023].

Picon, A. (2015) *Smart Cities: A Spatialised Intelligence*. Wiley.

Pileggi, S.F. (2016) 'Is Big Data the New? God? on Earth?', *IEEE Technology and Society Magazine*, 35(1): 18–20.

Portwood-Stacer, L. (2013) 'Media Refusal and Conspicuous Non-Consumption: The Performative and Political Dimensions of Facebook Abstention', *New Media & Society*, 15(7): 1041–1057.

Price, C. (2018) *How to Break Up with Your Phone*. Ten Speed Press.

Ragnedda, M. (2018) 'Conceptualizing Digital Capital', *Telematics and Informatics*, 35(8): 2366–2375.

Robinson, B.T. (2013) *Appletopia: Media Technology and the Religious Imagination of Steve Jobs*. Baylor University Press.

Roper, P. (2016) 'Dolmio Backs Up Prank By Putting the Pepper Hacker into Production', *Marketing Mag*, [online] 24 August, Available from: https://www.marketingmag.com.au/news-c/dolmio-pepper-hacker-production/ [Accessed 10 July 2023].

Schmitt, J.B., Breuer, J., and Wulf, T. (2021) 'From Cognitive Overload to Digital Detox: Psychological Implications of Telework During the COVID-19 Pandemic', *Computers in Human Behavior*, 124(106899): 1–8.

Segev, E. (2010) *Google and the Digital Divide*. Chandos Publishing.

Silk, M., Millington, B., Rich, E., and Bush, A. (2016) '(Re-)thinking Digital Leisure', *Leisure Studies*, 35(6), 712–723.

Smart, N. (2019) 'GLAMOUR Columnist, Niomi Smart, Shares the Mental Health Benefits of Taking a Digital Detox', *Glamour*, [online] 23 January, Available from: https://www.glamourmagazine.co.uk/article/digital-detox-benefits [Accessed 18 July 2023].

Solon, O. (2018) 'Apple's New "Digital Wellbeing" Tools Aim to Help Reduce Screen Time', *The Guardian*, [online] 4 June, Available from: https://www.theguardian.com/technology/2018/jun/04/apple-ios12-screen-time-apps-how-much-spent-phone-digital-wellbeing [Accessed 14 July 2023].

Statista (2018a) 'Number of Smartphone Users in France from 2015 to 2022 (in Millions)', *Statista*, [online] Available from: https://www.statista.com/statistics/467177/forecast-of-smartphone-users-in-france/ [Accessed 23 March 2023].

Statista (2018b) 'Number of Smartphone Users in Hong Kong from 2015 to 2022 (in Millions)', *Statista*, [online] Available from: https://www.statista.com/statistics/494594/smartphone-users-in-hong-kong/ [Accessed 10 July 2023].

Sutton, T. (2017) 'Disconnect to Reconnect: The Food/Technology Metaphor in Digital Detoxing', *First Monday*, 22(6).

Turkle, S. (2012) *Alone Together: Why We Expect More from Technology and Less from Each Other*. Basic Books.

Ugur, N.G. and Koc, T. (2015) 'Time for Digital Detox: Misuse of Mobile Technology and Phubbing', *Procedia – Social and Behavioral Sciences*, 195: 1022–1031.

Urry, J. (2007) *Mobilities*. Polity.

van Dijk, J.A.G.M. (2006) 'Digital Divide Research, Achievements and Shortcomings', *Poetics*, 34(4): 221–235. https://doi.org/https://doi.org/10.1016/j.poetic.2006.05.004

Veblen, T. (2005) *Conspicuous Consumption*. Penguin.

Vu, K. and Hartley, K. (2017) 'Promoting Smart Cities in Developing Countries: Policy Insights from Vietnam', *Telecommunications Policy*, 42: 845–859.

Williamson, B., Eynon, R., and Potter, J. (2020) 'Pandemic Politics, Pedagogies and Practices: Digital Technologies and Distance Education During the Coronavirus Emergency', *Learning, Media and Technology*, 45(2): 107–114.

Willis, S. and Tranter, B. (2016) 'Beyond the "Digital Divide": Internet Diffusion and Inequality in Australia', *Journal of Sociology*, 42(1): 43–59.

You Had to Be There (2019) *You Had to Be There*, [online] Available from: https://youhadtobethere.bigcartel.com/products [Accessed 14 July 2025].

Zahariades, D. (2018) *Digital Detox: The Ultimate Guide to Beating Technology Addiction, Cultivating Mindfulness, and Enjoying More Creativity, Inspiration, And Balance in Your Life!*. Independently Published.

Zuboff, S. (2019) *The Age of Surveillance Capitalism: The Fight for a Human Future at the New Frontier of Power*. Public Affairs.

6

Analogue Celebrity: Digital Refusal Among the Rich and Famous

A.R.E. Taylor and Neil Ewen

Introduction: Digital death

On 1 December 2010, US popstar Alicia Keys released a YouTube video titled 'Alicia Keys' Last Tweet and Testament'. In the video, she announced her 'digital death', stating that she was suspending her social media usage until $1million was raised for World AIDS Day. The money was to be donated to the Keep a Child Alive (KCA) foundation, a non-profit organization that Keys had co-founded with film producer Leigh Blake in 2003, which provides support services for HIV/AIDS-affected communities in India and Africa. Conjuring the imaginary division between the 'digital' and the 'real' she says: 'I have decided to sacrifice my *digital* life in order to give *real* life to people affected by HIV and AIDS'. Keys then performatively begs fans to buy her digital life back by donating quickly, saying that she already misses social media. Evoking familiar feelings of craving social media, Keys positions herself as ordinary and relatable. Several celebrities joined KCA's digital death campaign, including Lady Gaga, Elijah Wood, Kim Kardashian, Justin Timberlake, and Usher, each pledging to return online only when the funding goal had been achieved. The campaign was accompanied by striking images of these celebrities posing in coffins.

Underlying the KCA Digital Death campaign was the assumption that abstaining from social media could meaningfully be considered a challenging form of adversity or hardship suitable for celanthropic activity (akin to running a marathon for charity, for example). Another assumption was that the Tweets and Facebook posts of these celebrities were so valued that fans and followers would be willing to effectively pay for them (via a donation) to recommence. Ultimately, the campaign did not generate enough donations: and with only half of the funding goal reached after the first week of the campaign, pharmaceutical billionaire Stewart Rahr was asked to donate the remaining money to ensure the celebrities would meet their target. Usher had in fact arisen from his digital death before the target was met, adding to suspicions that the group of celebrity philanthropists had been disappointed by the public reaction. Perhaps social media users preferred Facebook and Twitter without them and realized that by *not* donating they could ensure the celebrities remained offline for as long as possible?[1]

By late 2010, social media had become a key tool through which celebrity was practiced. While the KCA campaign was largely a flop, it invites consideration of the role that the non-use of social media has come to play in the production of celebrity in the age of constant connectivity. While there exists a significant literature on social media and celebrity, comparatively little attention has been paid to celebrities who resist or refuse these platforms and how this stance relates to the construction of 'authentic' celebrity personae in a complex cultural moment defined at once by the ubiquity of social media and by an intensifying backlash against these platforms (see Thomas, 2014; York, 2018; Ingleton and York, 2019). The rejection of social media and, in some cases, the non-use of digital technologies more broadly, has played an important role in the construction of some major celebrity personae. For example, it has been widely reported that Keanu Reeves does not use social media, and online speculation even suggests that he may not own a computer. Margot Robbie describes herself as a 'very analogue person' in a December 2023 interview with CBS News, explaining that 'I do everything handwritten'. She discusses how she hand-wrote a letter to Quentin Tarantino to try and convince him to cast her in his 2019 film *Once Upon a Time in Hollywood*. She explains that she thought this approach would be more effective than an email because she knew that Tarantino is also 'a very analogue person', who 'only has a home phone' and she didn't know if he 'does' email (CBS News, 2023). Cillian Murphy's disengagement with digital culture has been widely discussed after interview clips went viral in which he revealed his lack of interest in social media, stating he is 'too old for that'. Commentators lauded the actor's avoidance of social media as a praiseworthy example of self-discipline and restraint (The Minimalists, 2023). Murphy also apparently adopts a minimalist approach to his smartphone, with an all-black wallpaper, few apps installed, and notifications disabled (Outstanding Screenplays, 2024). Many other stars reportedly do not use social media – at least in the form of a public-facing account. Sandra Bullock, Daniel Craig, Scarlett Johansson, Kate Moss, George Clooney, Kate Winslet, Olivia Wilde, Renée Zellweger, Jennifer Lawrence, Keira Knightley, Rachel McAdams, Mary-Kate and Ashley Olsen, Emily Blunt, Emma Stone,

Kristen Stewart, Julia Roberts, Mila Kunis, Daniel Radcliffe, Tina Fey, Cate Blanchett, Saoirse Ronan, Chris Pine, Phoebe Waller-Bridge, Theo James, and Alicia Vikander are just some of the other famous names who ostensibly do not have social media accounts.

Celebrities who refuse to use social media are of different ages and their fame is aligned with different media forms, though this trend is particularly notable in the film industry. Some even shun specific digital technologies altogether, such as film director Christopher Nolan, who, according to some interviews, does not use email or smartphones (Riley, 2012a, 2012b; Smith, 2020). Nolan is also an ambassador of analogue film, often opting for 35mm or 70mm film stock instead of digital video, and using analogue production practices where possible (such as editing sequences from the original negative or using photochemical colour timing). Woody Harrelson has apparently not used a mobile phone since 2021 (Watts, 2024), stating in a June 2024 episode of his podcast, 'I don't like to be readily available to any human being at any time'. This statement quickly went viral, with images of the quote widely circulating across social media. Queen Margrethe II of Denmark is famous among Danes for 'her outspoken rejection of mobile phones and the internet' (Gozzi, 2024). Many more celebrities have decided to temporarily abstain or quit social media, often to strategically return online a few days, weeks, or months later, to a frenzy of fanfare – usually conveniently timed around the launch of a new product release. This practice has become increasingly fashionable amid the wider cultural backlash against big tech companies (the 'techlash') since the late 2010s (Seymour, 2019; Zuboff, 2019; Chia, Jorge, and Karppi, 2021).

Despite growing disillusionment, these platforms remain a structuring presence in everyday life. The act of refusing to use social media still draws considerable public and press attention, as testified by the virality of clips and the regular news articles dedicated to various practices of analogue celebrity. Listicles, such as '21 celebrities who don't use social media' (Bahou, 2021) or '29 Celebrities Who Reject Social Media & Why' (Sprankles and Grey, 2023) are now a staple of celebrity news and gossip columns. In the celebrity press, and across fan forums and Reddit posts, a celebrity's lack of social media presence or disengagement from digital culture is represented in myriad ways: as abnormal or deviant; as endearing; as subversive; as a form of technophobia or luddism; as an admirable demonstration of discipline and self-control; as a respected political stance towards media platforms that are increasingly perceived as negative or harmful forces.

Taking this interest in celebrity social media refusal as its starting point, this chapter introduces the term 'analogue celebrity' to describe a broad spectrum of disconnective positions and practices that celebrities engage in or adopt in relation to digital technologies. At a conceptual level, analogue celebrity is related to, but distinct from, performances of 'celebrity reluctance', theorized by Pamela Ingleton and Lorraine York (2019; see also York, 2018).

Ingleton and York develop an analysis of the 'reluctant celebrity' through an exploration of tweets in which celebrities express their reluctance to engage with social media. Reluctant celebrity is 'a simultaneous, double-facing mode of celebrity performance in which the celebrity is disinclined to perform, yet performs nevertheless' (Ingleton and York, 2019: 367). In contrast to reluctance, which 'exists at, but never crosses, the threshold of withdrawal' (Ingleton and York, 2019: 367), analogue celebrities are those who refuse to use digital platforms, or who have crossed that threshold and withdrawn (though sometimes only temporarily) from these platforms. Performances of analogue celebrity provide a generative opportunity for exploring celebrity self-fashioning, power, and privilege in post-digital culture.

With digital technologies and social media now pervasive forces in everyday life, a diverse body of interdisciplinary literature has critically examined varying practices and expressions of digital technology avoidance, including social media refusal (Wyatt, Thomas, and Terranova, 2002; Selwyn, 2006; Bauerlein, 2011; Helsper and Reisdorf, 2013; Portwood-Stacer, 2013; Hardey and Atkinson, 2018; Syvertsen, 2020; Chia, Jorge, and Karppi, 2021). Celebrity practices of digital refusal can range in intensity, from taking the occasional social media hiatus to the more extreme example of refusing to use email or mobile phones. Through our use of the term 'analogue celebrity' we do not seek to reinforce or reify rigid binaries between the 'analogue' and the 'digital' or between technology 'use' and 'non-use'. We recognize that celebrity disengagement from digital technology 'is rarely total, but often situational' (Hesselberth, 2018: 1997) and can be related to a specific medium, time, or place (Keanu Reeves, for example, does not have his own social media accounts but still often appears in TikToks promoting his band, Dogstar, and is therefore no stranger to leveraging the promotional affordances of social media). As such, we conceptualize analogue celebrity less as a distinct typology and more as a practice that unfolds across a spectrum of positions in relation to digital technologies and platforms, with the aim of conceptualizing digital refusal and, more broadly, technology non-use, in more nuanced terms (Wyatt, Thomas, and Terranova, 2002; Wyatt, 2003; Satchell and Dourish, 2009; Baumer et al, 2014).[2] We draw on press articles and social media comments to develop two case studies, the first focusing on Keanu Reeves and Renée Zellweger in relation to the promotion of their 2016 film *The Whole Truth*, and the second on Ed Sheeran. The different forms of social media non-usage among these famous figures provide entry-points for sketching two practices of analogue celebrity: Reeves and Zellweger as 'digital refusers', and Sheeran as a 'digital detoxer'. These categories are necessarily unstable and ever-changing but productive in delineating differences in celebrity social media non-usage. Prior to the case studies, we begin with a brief overview of the literature exploring celebrities and social media and explore two key affordances of social media

– accessibility and authenticity – through which celebrities engage their fanbases and maintain their fame, so that we may better understand how practices of digital refusal and detox rework and reconfigure (rather than reject) narratives, performances, and experiences of celebrity authenticity.

Social media and celebrity: the promise of access and authenticity

Within the field of celebrity studies, a considerable body of work has explored new forms of fame that have coalesced around social media, and the various ways that famous people, from film stars to politicians, strategically engage with these platforms (Marwick and boyd, 2011; Marwick, 2013; Gaden and Dumitrica, 2014). The years between 2005 and 2015 marked something of a golden era of social media usage. Millions of people around the world, including celebrities, opened accounts on Facebook, Twitter (now 'X'), Instagram, and other social platforms. 'Everyone, we are told', Jonathan Crary observed in 2013, 'needs an "online presence", needs 24/7 exposure, to avoid social irrelevance or professional failure' (Crary, 2013: 104; see also Rosamond, 2017). Having a social media presence quickly became the norm, and a valuable component of the celebrity toolkit.[3] As well as providing established celebrities with a powerful vehicle through which they could consolidate their fame, these platforms also facilitated the rise of new forms of social media-based 'micro-celebrity' (Senft, 2008; Gräve, 2017; Abidin, 2018), such as the 'influencer' and the 'YouTube star'. Agents and publicists trumpeted the importance of social media as a platform for increasing a celebrity's public reach. Expectations and obligations for public figures to regularly post content on social media – especially those without superstar status – quickly became part of everyday celebrity labour. 'It is not uncommon for artists to flit across Twitter, Facebook, Instagram, YouTube, Vine, Line, Snapchat and Periscope', Guardian journalist Eamonn Forde noted in 2015, 'with much of their day taken up in trying to keep these promotional plates spinning' (Forde, 2015). Many famous figures adapted to the changing requirements of celebrity culture by providing increased access to their ostensibly 'authentic' selves via social media, capitalizing on the new forms of intimacy and accessibility these platforms afforded.

The act of posting on social media enables celebrities to communicate directly with their fans, bypassing formal access brokers, such as public relations professionals, agents, managers, or other 'cultural intermediaries' (Rojek, 2012: 27) that traditionally manage or control a celebrity's public communications (see also Gamson, 1994: 79–80), producing a sense of unmediated access. Alice Marwick and danah boyd (2011: 139) note in their study of celebrity Twitter usage: 'celebrity is practiced through the appearance and performance of "backstage" access'. While media technologies have always provided fans with a certain degree of backstage access to celebrities, social media increased the potential for accessibility exponentially. Its rise unfolded alongside an intensification of cultural interest in the private (rather than professional) lives of the rich and famous (Turner, 2004: 4). Hannah Hamad (2018: 46) notes that the new media forms associated with Web 2.0 produced 'skyrocketing levels of accessibility, visibility and speed of circulation to and of celebrity imagery and discourse'. This accessibility can play an important role in diminishing felt distances between audiences and celebrities, bringing a famous figure 'closer to home' through regular content and updates posted throughout the day. Stars like Sylvester Stallone, Justin Bieber, Dwayne 'The Rock' Johnson, and Cristiano Ronaldo regularly post on their social media directly from their mansions and workplaces (while on set, on tour, in the locker room, or simply eating a meal), providing fans with what feels like a behind-the-scenes glimpse at supposedly un-staged private lives. In contrast to earlier forms of media that granted audiences 'behind-the-scenes' access to stars (such as interviews, appearances on talk shows, or DVD extras), social media is often perceived as providing more immediate access, sometimes even real-time access in the case of the live-streaming options that these platforms offer. Social media content from celebrities thus often feels direct and uncensored. This perceived opportunity for direct and real-time interaction has enabled new para-social forms of 'digital intimacy' to emerge between celebrities and their fans (Thompson, 2008; Reade, 2020).

The rise of social media has unfolded alongside a cultural fascination with the 'authentic' (Syvertsen and Enli, 2020; Heřmanová, Skey, and Thurnell-Read, 2023). Authenticity is often colloquially used as a synonym for 'honesty', 'integrity', 'sincerity', 'genuine', 'true', or 'real'. In popular culture, authenticity remains a celebrated and revered quality of a person, place, or product – even if the term has long been co-opted by marketing agencies, the tourism industry, politics, branding, and advertising (MacCannell, 1973; Cohen, 1988; Peterson, 1999; Banet-Weiser, 2012; Enli, 2015). Despite the commercialization and commodification of authenticity, the concept still holds cultural value. Social media can provide celebrities with opportunities to construct a presentation of self that seems 'authentic' because it appears un-staged, private, intimate, and 'stripped of PR artifice and management' (Marwick and boyd, 2011: 149). By posting photos of their private lives and by appearing to respond personally to comments, celebrities (and politicians) use social media to cultivate a sense of an authentic personal connection with their fans and followers (Dare-Edwards, 2014; Gaden and Dumitrica, 2014).

Cultural theorists have tended to approach claims of authenticity with scepticism, preferring to emphasize the socially constructed, dynamic, and context-dependent nature of authenticity discourses (Trilling, 1972; Bendix, 1997; Cheng, 2004). This has been particularly true in analyses of social media, where scholars often foreground the constructed and performative nature of evocations of authenticity by using various qualifiers including 'staged authenticity' (Pooley, 2010), 'contrived authenticity' (Abidin, 2017), or 'mediated authenticity' (Enli, 2015; Frowijn, Harbers, and Broersma, 2023). Marwick and boyd (2011) note that celebrity Twitter users 'reveal what appears to be personal information to create a sense of intimacy between participant and follower [and] give the impression of candid, uncensored looks at the people behind the personas' (Marwick and boyd, 2011: 139). 'Appears' and 'impression' are the operative words here. Critical theorists remind us that authenticity is never an inherent quality. Despite the apparent promise of un-staged, unscripted, or unedited access, a celebrity's self-presentation within 'behind-the-scenes' media products is carefully crafted and managed by the celebrity and their public relations teams. While social media may promise to provide direct access to a celebrity who appears genuine and authentic, celebrity self-presentation on social media is often carefully constructed to be consumed by others. Celebrities must thus engage in ongoing forms of 'impression management' or 'authenticity labour' (McRae, 2017: 14; Maares, Banjac, and Hanusch, 2021) to develop a presentation of self that appears both believable and genuine. If a celebrity's social media account is found to be managed by a PR team, fans often do not accept the posts as authentic self-expression. For example, regular debates have unfolded around Britney Spears' social media usage. During the period of her conservatorship, Spears repeatedly stated that her personal Instagram was one area of her life over which she had control (Access Hollywood, 2019; ET Live, 2019). However, she has since been open about the fact that she has a dedicated social media team, which often leads to questions about the authenticity of her content. Celebrities and their publicists must tread carefully, because 'deliberately managing one's impression to be perceived as authentic is commonly presented as the direct antithesis of being truly authentic' (Frowijn, Harbers, and Broersma, 2023: 219).

Authenticity is often performed on social media through revelatory acts of self-disclosure. The sense of intimacy and direct access afforded by these platforms has enabled social media to become key sites where celebrities disclose private experiences, make public confessions, or apologize for inappropriate behaviour. Since the late-twentieth century, the construction, performance, and maintenance of celebrity authenticity have been increasingly tied to practices of self-disclosure and confession. According to Sean Redmond (2010), the confessional is any given moment in which a famous person engages in a revelatory act understood as heartfelt and honest. Disclosing information that is private or difficult to discuss, or confessing to difficulties or errors, is a key means through which a celebrity's authenticity can be established (Muntean and Petersen, 2009). Talk shows, interviews, documentaries, and social media are all arenas through which celebrities position themselves as 'authentic' through acts of confession and public apology. Social media thus occupies a curious position in contemporary popular culture. These platforms are at once perceived to provide opportunities for individuals to appear genuine, honest, and authentic but, at the same time, widely associated with fake, filtered, and staged content.

In the following case studies we introduce two examples of analogue celebrity positioning, exploring how social media non-use reconfigures the accessibility of the celebrities and the construction of their authenticity: 1) Keanu Reeves and Renée Zellweger are examples of 'digital refusers' whose permanent lack of social media harks back nostalgically to a golden era of celebrity when the private lives of stars was shrouded in mystery; and 2) Ed Sheeran as an example of a 'digital detoxer' who engages in temporary breaks from social media for wellbeing reasons that often also serve as valuable PR opportunities. Through this discussion we explore how celebrities cultivate authenticity through their rejection of social media. Lack of social media presence, we argue, has become a key way through which a celebrity's authenticity can be demonstrated in the age of the techlash. This analysis ultimately reveals the seemingly endless flexibility of the concept of 'authenticity' in relation to social media platforms. Indeed, a celebrity can use social media to communicate their authenticity, but authenticity can also be communicated through a rejection of these platforms.

Digital refusers: Keanu Reeves and Renée Zellweger

January 2022 began with fervour and speculation in the world of social media celebrity. Could it be true that Keanu Reeves, who had famously shunned social media, had in fact now joined TikTok? A five-second video titled 'Welcome to my TikTok', posted on the 18 January 2022, appeared to show Reeves, sitting on a sofa in a John Wick-style black suite, looking at his smartphone. In the comments, a number of users welcomed Reeves to TikTok. Many, however, were sceptical. Upon closer inspection, some noticed small details about Reeves that looked slightly strange – from his fingernails to the unusual eeriness of his smile. The name of the account itself was a giveaway: unreal_keanu. It turned out to be a hoax: an AI parody account using deepfake technology. Subsequently, the account released a steady stream of videos showing Reeves doing a range of (often Matrix-themed) activities and poses, quickly amassing millions of followers.

These videos re-ignited interest in the Hollywood actor's curious absence from social media. In the middle-aged period of his career, Reeves' celebrity has been the subject of fascination.[4] Notably, Reeves' non-use of social media has become an integral aspect of his star persona. Online message boards are rife with fan reflections and speculation about why Reeves does not have any public-facing social media presence. In a 2019 thread dedicated to the subject on Quora, a social question-and-answer site, one user proffers that social media is 'a cesspool of self-serving bullshit to jack-off one's ego. [Reeves] seems like a genuine guy so that doesn't seem to appeal. Plus he's already famous so he doesn't need to self report. I also read somewhere (like 15 years ago) that he claimed he doesn't even have a computer' (Quora, 2019).

The user presents social media as a vacuous arena of self-serving marketing and promotion that Reeves does not need to participate in because he is 'genuine' and 'already famous'. While social media can provide celebrities with a platform to carefully construct a persona that appears authentic, avoidance of social media is here presented as an equally important index of authenticity. Celebrities without social accounts, the user suggests, exist beyond the shallow pursuit of self-aggrandizing promotion. This response aligns with a larger cultural perception of Reeves as an honest, humble, and down-to-earth celebrity – a perception that has emerged in response to several widely reported anecdotal acts of kindness that Reeves has carried out (Horeck, 2022). The user's comment also highlights that, unlike social media influencers who rely on these platforms for their fame, Reeves' fame pre-dates social media and, as such, he does not need these platforms to further amplify his fame or boost his career. Similar sentiments are echoed by another user, who asks, 'Why should he be on social media? He doesn't need exposure or advertising. His name and talent for his craft are more than enough' (Quora, 2019). For this user, Reeves does not need to stoop to the level of social media self-promotion because his acting talent is enough to maintain his fame. The star is presented as a talented craftsman dedicated to his art.

A key reference point in discussions about Reeves' lack of social media presence is a 2016 interview conducted by *Entertainment Tonight* promoting the film *The Whole Truth* (2016) with Reeves and co-star Renée Zellweger (Murphy, 2016). Like Reeves, Zellweger does not use social media and has discussed her stance towards these platforms in various interviews. A voiceover cutaway during the 2016 interview highlights that 'neither of the A-listers have [social media] accounts connecting them with their fans', which segues into a question from Deidre Behar (the journalist conducting the interview), about the reasons behind this. Behar asks: 'Can we get you on there? We wanna see the behind-the-scenes!', to which Reeves replies: 'For me, I guess it's privacy … it is important to me'. By refusing to use social media, Reeves places a limit on his accessibility and is able to exercise a modicum of control over how much of his private life is available for public consumption. But his stance towards social media also generates media interest, ensuring he continues to attract attention.

Zellweger responds to the interviewer's question by harking back to a mythical era of stardom where stars' private lives were ostensibly less accessible, enabling celebrities to cultivate a sense of mystery and intrigue.

Foregrounding the aura of mystery that a lack of social media presence can establish she replies: 'don't you think a little mystery is a good idea?'[5] Early theorists of celebrity culture suggested that film stars (particularly those associated with the 'golden era' of Hollywood) were defined by their distance from 'ordinary' people (as the 'star' metaphor captures). 'They live at a distance, far beyond all mortals', sociologist Edgar Morin observed in his 1960 analysis of stardom (Morin, 2005 [1960]: 9). This distance is often spatially circumscribed (stars live in gated compounds reserved for the elite, for example), but it is also social, cultural, and economic (Rojek, 2015: 115). Media scholar Jib Fowles (1992) used the phrase 'star village' to describe this elite celebrity realm that is both metaphorically and materially distant and inaccessible to most people. Ellis Cashmore (2006: 19), writing specifically about Hollywood stars, observes that these individuals appeared as if they

> were parts of the pantheon: like deities, they seemed to exist at a level above that of other mortals. They lived lives of such opulence, such splendour, such sublime beauty that they seemed unapproachable. And, in a genuine sense they were … They secured themselves away and drip-fed their fans with occasional personal appearances and carefully controlled silvery images.

The distance between stars and the public was carefully navigated. Access to stars was limited and largely relied on media professionals for its facilitation through talk shows, magazine interviews, or photo shoot releases. Of course, even golden age stars used the media (from talk show interviews to staged paparazzi shots) to strategically share parts of their private life, granting audiences a supposed glimpse of the person 'behind' the performances. Today, the act of posting on social media intensifies this logic, and, in doing so, Zellweger suggests, reduces any 'mystery' surrounding a celebrity's private life.

By posting mundane food photos, live-streaming from their houses, or sharing make-up-less selfies, celebrities can strategically appear like 'normal' people doing 'ordinary' things.[6] Such posts can cultivate a sense of intimacy, closeness, and familiarity between a celebrity and their followers, enabling famous figures to 'transition from distant star to accessible celebrity' (Hamad, 2018: 52). By contrast, Zellweger suggests, lack of social media presence can enable celebrities to maintain the mysterious aura of a distant star, nostalgically harking back to an earlier form of pre-digital fame where interactions with stars, and information about them, was less widely accessible, a throwback to an older era of celebrity in which access to the famous was more limited. As celebrity studies scholar Sarah Thomas (2014: 245) notes, social media refusal enables some of the biggest modern stars to 'maintain the traditional classical star's aura of distance'. Thomas (2014: 245) suggests that lack of social media 'helps these phenomenally successful stars to remain elusive and extraordinary, "knowable" only through gossip and traditional mediation'. Mystery, distance, and allure are valuable tools through which fascination and interest in a celebrity can be maintained. In an era of social media saturation, refusing to have a social media presence does not only generate mystery but also intense interest as to the reasons underlying this decision, ensuring enduring press attention. As such, practices of using or not using social media among celebrities must not be seen as political opposites, but as the heads and tails of the wider attention economy within which the famous necessarily operate.

In the interview, Zellweger states that 'most [social media content] is really unimportant and unnecessary and makes a person's life way more interesting and colourful than it is in truth'. These comments reproduce popular critiques about the triviality of social media status updates and the often staged and untruthful performances of self that take place on these platforms. Zellweger suggests that social media is a site of inauthenticity, untruth, and manipulation, where appearances are contrived and deceitful. By rejecting social media, she is seen to shun the superficial world of online impression management and attention-seeking.[7]

Zellweger's digital refusal extends beyond social media, to other digital devices. In a 2022 interview for *Harper's Bazaar*, she stated that she tries to limit her screen time and avoids checking her mobile phone until the evening (Rapkin, 2022). In many popular narratives of digital refusal, the rejection of digital technologies often goes hand-in-hand with a celebration of the outdoors. In an unsurprising evocation of the analogue idyll, Zellweger explains to *Harper's Bazaar* that she spent the bulk of her time during the COVID-19 lockdowns, 'outside every day, building things and planting things' (Rapkin, 2022). Here the privilege of digital refusal is strikingly clear. During the COVID-19 pandemic, while Zellweger was able to avoid the digital world, many people (especially those from lower socio-economic backgrounds without access to green spaces) were stuck indoors and unable to disconnect from their digital devices, having to working remotely via Zoom, and finding escape in binge-watching Netflix or doom-scrolling through Instagram (Kuntsman, Martin, and Miyake, 2023). A degree of privilege thus underpins the capacity of a celebrity to opt out of digital culture. Celebrities who were famous before the social media era, such as Reeves and Zellweger, can afford to avoid these platforms. Their fame is not tied to social media, and they can attract media attention without needing to engage with these platforms. This is not the case for non-'A-list' celebrities – particularly 'celetoids' (Rojek, 2012) – many of whom rely on social media to find work, promote themselves, and remain in the public eye. An analogue/digital divide emerges within celebrity culture between those who do not need to use social media and those who cannot completely abandon these platforms because they rely on them to make a living (O'Connor, this volume).

Taking breaks from social media

In contrast to celebrities like Reeves and Zellweger, who have refused to engage with social media from the outset, others have decided to take temporary breaks from their social media activity for various reasons. Taking momentary social media breaks or 'digital detoxes' has become especially fashionable among social media influencers (Jorge and Pedroni, 2021), who depend on these platforms for their livelihood and who are thus well positioned to understand the stress and pressure that comes with the need to constantly maintain an online presence (Glatt, 2022, 2023).[8] But beyond social media influencers, digital detoxing has also become popular among 'traditional' celebrities. Unlike influencers, these celebrities can afford to take extended time away from social media because their fame has never been (or is no longer) dependent on these platforms. Kanye West, Selena Gomez, Kendall Jenner, Adele, Zayn Malik, Stephen Fry, Rylan Clark, Thierry Henry, Gypsy Rose Blanchard, and Tom Holland are celebrities who have taken hiatuses from social media. The reasons underpinning their breaks differ, from the need to improve mental health (by temporarily escaping the toxic and hateful comments of online trolls), to attempts to avoid distraction (in the hope of increasing productivity and focus), or to safeguard their brand if they have posted problematic content that threatens to get them 'cancelled'. In 2021, Adele abandoned social media because she found it distracting. 'I just would find myself not getting my errands done for the day', she explained in an interview with *The Face*. 'It's *made* to be addictive' (Carty-Williams, 2021, emphasis in original). Adele's comments reproduce tropes of social media addiction and distraction; tropes that have become a central narrative element of anti-social media sentiment. Adele is widely known for her sometimes insensitive posts on social media and, for this reason, in the past her PR team have changed her passwords or restricted her usage (Lasimone, 2021; Oganesyan, 2021). For some, Adele's passionate and volatile posts are evidence of her authenticity, reflecting her inability to conform to the staged, edited, and contrived requirements of celebrity social media performance in the cancel culture era. As journalist Rebecca Reid (2020) states: 'Adele is too real for Instagram, too gobby for Twitter, and too talented to need social media at all.' By describing Adele as 'too real for Instagram', Reid seems to suggest the star has an irrepressible authenticity (evidenced by her controversial posts) that is incompatible with the contrived falsity of social media. For Reid, there also appears to be an imagined threshold of fame or talent beyond which a social media presence is not needed.

Several other celebrities have taken breaks from social media after posting off-brand content. Indeed, if not used strategically, social media can be a minefield for celebrities, with an insensitive post potentially exposing them to public criticism which may lead to their 'cancellation'. Ten Walls, J.K. Rowling, Elon Musk, Katie Hopkins, Azealia Banks, and David Eason have all been widely criticized for controversial comments they have posted on social media. Other celebrities take breaks from social media to escape hateful trolls. Comedian and author Stephen Fry regularly quits social media in response to comments his posts attract. In 2009, he threatened to leave Twitter because there was 'too much aggression and unkindness around' (Press Association, 2015). In 2015, he also temporarily shut down his Twitter account and logged out of Instagram (see Kennedy, 2015). Twitter has been singled out by numerous stars as a platform that attracts abusive trolls. Other celebrities who have left Twitter after suffering abuse – dubbed the 'Twitter Quitters' by the popular press (Savage, 2017) – include Miley Cyrus, Sue Perkins, Halsey and *Avengers* director Joss Whedon.

These celebrity breaks from social media have occurred against the backdrop of a growing discourse of disillusionment with these platforms that has gained cultural traction since the mid-2010s (Chia et al, 2021). Moral panics have proliferated in the popular press around a range of concerns related to social media. Addictive design mechanics have been blamed for enticing users to spend hours of their time endlessly scrolling through feeds (Seymour, 2019). Social media content has been blamed for distracting users, leading to a loss of productivity and dwindling attention spans. The algorithmic recommendation systems that govern social media networks have been blamed for creating 'filter bubbles' that transform these platforms into breeding grounds for online hate where 'toxic' viewpoints proliferate, 'radicalizing' users. There are growing public and political debates about the harmful impact of social media on mental health, with concerns arising around body dysmorphia, social comparison, or exposure to harmful content. Like many previous media technologies that have been identified as sources of potential social harm, from television to videogames, social media has been decried as unsafe, dangerous, toxic, a threat to liberal democracy and even human life. This ongoing backlash against the big tech companies responsible for social media has been termed the 'techlash' (which was shortlisted for the Oxford English Dictionary's 2018 word of the year).

Within this context, it has become increasingly common (and fashionable) for celebrities to publicly announce that they are taking a break or 'detox' from social media. A celebrity's break from social media is typically announced *on* social media. Often these announcements take a confessional form, providing celebrities with a valuable opportunity to demonstrate their ordinariness, normality, and authenticity by drawing attention to their relatable struggles with social media addiction, distraction, or pressure – revealing to their fans and followers that they wrestle with the same problems that everyday social media users experience. Laura Portwood-Stacer (2013: 1041) has noted that social media abstention is 'a performative mode of resistance, which must be understood within the context of a neoliberal consumer culture, in which subjects are empowered to act through consumption choices – or in this case non-consumption choices – and through the public display of those choices'. One celebrity who regularly uses performative social media breaks and 'returns' to construct his authenticity through confessional acts (and, in the process, to strategically build hype for new album releases) is Ed Sheeran.

The digital detoxer: Ed Sheeran

Ed Sheeran's career is heavily entangled with social media. With approximately 16 million Twitter followers and 49 million subscribers on Instagram, he is one of the most avidly monitored celebrities on social media. His rise to fame was largely facilitated by social media, in the form of the YouTube channel SB.TV, launched by the late British music entrepreneur and DJ Jamal Edwards and now an online R&B/Hip-Hop platform. In the early years of his career, Sheeran released several videos through SB.TV and would regularly spend time on Twitter in live Q&As to help build his fanbase. Part of Sheeran's success is due to his strategic leveraging of social media to promote his work. Instead of using traditional marketing he has typically opted to release teaser posts, videos, and clips of forthcoming singles on Twitter, Snapchat, and Instagram, using social media to directly respond to fans with the aim of building a personal connection.

Through his social media marketing approach, Sheeran has generated hype and anticipation for his music. Concurrently, he began to take regular, highly publicized, breaks from social media and other digital devices.[9] In December 2015, at the end of a world tour promoting his second album, he announced on Instagram that he would be taking time-out from his 'phone, emails and all social media', returning in autumn 2016 (teddysphotos, 2015). In this post he stated that he had found himself 'seeing the world through a screen' and suggested that he was going offline to experience unmediated reality. Sheeran's announcement generated a frenzy of news coverage, with some journalists speculating that this could mark the beginning of a mass celebrity exodus from social media (Kennedy, 2015). A *Guardian* article headlined 'Post-Internet: Has the Bubble Burst on Pop's Social Media Strategy?', suggested that Sheeran's decision was reflective of 'a growing dissatisfaction among pop stars with their social media obligations' and that he may 'inspire others to finally break from the "always on" culture' (Forde, 2015). In a blurring of the rural idyll and the analogue idyll, the article closes by drawing a connection between the countryside and the offline:

> Towards the end of the 1960s, many acts pulled themselves out of the bacchanalian merry-go-round that was London in order to 'get it together in the country'. Ed Sheeran could now inspire the current generation of pop stars to cut free of their marketing moorings and get it together offline. (Forde, 2015)

One year later, Sheeran returned to social media with a video released on Twitter in which he held up a decidedly analogue piece of paper with the handwritten message: 'New music coming Friday!!' (teddysphotos, 2017). Press coverage quickly proliferated (Robinson, 2016). Sheeran clearly understood how to strategically leverage social media abstention to generate publicity and build hype. Since then, he has continued this strategy, taking breaks from social media only to end his digital silence with the announcement of a new album, adding further hype and promotional value to the release.

In a 2017 interview with the UK tabloid *The Sun*, headlined 'Sheeran Twitter Quitter', he announced that he had stopped using Twitter: 'I've actually come off Twitter completely. I can't read it. I go on it and there's nothing but people saying mean things' (Wootton, 2017). Mobilizing wider narratives about social media platforms as harmful spaces where online hate speech and cyberbullying proliferate, Sheeran aligned his brand in opposition to the toxic culture that was increasingly associated with social media. Sheeran's social media detoxes also extended to other digital technologies. In a June 2017 carpool karaoke with James Corden, he claimed that he no longer owned a mobile phone (*The Late Late Show with James Corden*, 2017). Later, in December 2019, he announced he was taking 'a breather' from social media and returned in September 2020 to announce the birth of his daughter (teddysphotos, 2019). In November 2022, he undertook a further break. A post that showed him standing on a boulder, staring pensively out to sea, was captioned: 'Signing off now until 2023. See you in the new year. Much love, Ed x' (teddysphotos, 2022). This break ended on 19 January 2023, when he returned online to post the video for his latest single 'F64' (teddysphotos, 2023a).

Sheeran's return to social media became a flashpoint in early 2023, with online news outlets all covering his 'social media comeback' (Gurley, 2023). Just over a week after his new single was released, he posted a confessional video on Instagram, formally announcing his return and apologizing for his social media abstention. In the video, titled 'Back in the bitz', he explains that he had taken time off from social media due to 'turbulent things' occurring in his personal life, including the death of his friend Jamal Edwards, the founder of SB.TV (teddysphotos, 2023b). He continues, 'I just didn't really feel like being online and pretending to be something I'm not when I wasn't feeling like that', and explains that he wanted to be 'totally honest' with fans. Sheeran's narrative alludes to wider popular perceptions that social media platforms are cultural sites where people present carefully staged constructions of a happy, perfect self. It is, of course, understandable that he may not have wanted to post content during this difficult time. However, there is no reason why 'honest' content could not be posted about the difficulties he was encountering; social media does not have to remain a space for staged happiness and perfect worlds. The confessional video provided Sheeran with a platform to demonstrate his honest and authentic persona, enabling him to strengthen relations of intimacy and authenticity with his fans.

The post provoked a multitude of heartfelt comments from fans. For one fan (iamjujutakano), Sheeran's post was a stark reminder that celebrities are subject to the same struggles as 'ordinary' people: 'People tend to forget that artists, celebrities, famous people are human beings just like them. Everyone has ups and downs Ed! Shit happens to us all'.

This sentiment was echoed in other comments. One user, melasimpson, highlighted that

> There are shedloads of musicians, singers, actors etc who have their 'people' handle all the fan media. You've spoiled us mate! But we don't 'own' you. I'm happy to have the music and blessed to see your posts, but the very fact that you live your life like a 'normal geezer' with ups and downs, like us all, is the one thing that makes you the man we love.

For this user, Sheeran's open discussion of his struggles with social media signalled his normality and his dedication to his fans. This relatability can work to establish authenticity, where Sheeran appears genuine and honest. This was suggested by one user (adashorwhatever) who said: 'don't ever feel like you have to post anything for us that isn't the real you'. An Instagram fan page (teddyed_11) dedicated to Sheeran celebrated his return and praised Sheeran for running his account himself, rather than relying on a social media PR team: 'We don't care how often you post Ed. We just care that it is actually you when you do. Even if that means once a month or less. ❤'

This comment suggests that content posted by a celebrity themselves is more important than a regular stream of content posted by a PR team. One fan (pattyjeank) highlights that, given the level of fame Sheeran has now reached, he no longer needs to feel the pressure to engage in social media labour: 'You've moved way beyond those days on Twitter when you'd do an impromptu Q&A'. In a YouTube reposting of the video by *Entertainment Tonight* (2023), a fan (EmilyJBunny) similarly states that 'When you're Ed Sheeran you don't NEED a social media presence. Do what you like Ed, we are here for whatever you got!' One comment from jld4780 praised Sheeran's authenticity, stating that, unlike other celebrities on social media, he hadn't got 'stuck in the game of pride'. Another YouTube user (graceskyephoto) sympathized, stating that 'Sometimes everyone just needs to take a social media break. It's good for the soul'.

High-profile social media breaks present celebrities like Sheeran with opportunities to performatively construct their authenticity and ordinariness by demonstrating that they grapple with the same problems that everyday social media users encounter. Through confessional acts of self-disclosure about the toxicity of social media and the mental health difficulties that arise around the pressures of maintaining an online presence, celebrities unite themselves with their followers in their common struggle with these platforms. Momentarily opting out of social media thus enables celebrities to 'construct their images as "real" people, subject to stress, overexposure, and abuse of digital devices' (Jorge and Pedroni, 2021: 75). Celebrity digital detox narratives often 'allude to a wider lexicon of social media toxicity' (Jorge and Pedroni, 2021: 70; see also Sutton, 2017; Syvertsen, 2020) and emphasize the importance of self-care, with celebrity digital detoxers serving as exemplars of the neoliberal injunction that individuals, rather than big tech companies or governments, should take responsibility for health and wellbeing issues related to social media. Celebrity digital detox posts provide fans and followers with a neoliberal narrative of self-regulation via confessional self-disclosures where an honest and vulnerable authentic celebrity self is supposedly revealed. Sheeran's breaks from social media must thus be seen within the wider context of the 'self-regulation society' (Syvertsen and Enli, 2020: 1270) that has emerged under the conditions of neoliberal capitalism, in which individuals are encouraged to practice self-control and self-care, often by turning to market solutions in the form of self-help guides or an ever-growing range of digital disconnection consumer products (such as lock boxes for mobile devices [Fast and Syvertsen, 2024] or expensive digital detox retreats [Fish, 2017; Sutton, 2020]). Amid the backlash against big tech, disconnecting from digital technologies has become a fashionable trend and marketing opportunity, energizing an entire economy of detox holiday packages, self-help products, and wellbeing courses through which users are enjoined to take responsibility for managing their overconnectivity (Kuntsman and Miyake, 2019; Syvertsen, 2020).

Sheeran's social media breaks enable him to appeal to his fanbase by presenting himself as a 'normal geezer' who struggles with social media like ordinary people. These breaks are often strategically timed to build up to a new release and to produce hype and anticipation, to the point that it is not only the release of new music that becomes a flashpoint but also his repeated returns to social media. Through highly publicized digital detoxes, celebrities like Sheeran come to embody, reproduce, model, and perpetuate neoliberal ideologies of self-control and self-help. Sheeran thus arises as a good neoliberal subject, taking self-imposed breaks from social media without ever fully abandoning the platforms, in order to support his productivity and creativity, while simultaneously constructing an image of authenticity, building hype for his return online and, of course, promoting his music (generating value for his brand, his record label, and for social media platforms when his return sparks a viral frenzy of activity).

Conclusion

Social media has long been theorized as a valuable site for the production of celebrity authenticity and remains a key tool through which celebrities strive to construct and stage authentic selves by providing fans with glimpses into a supposedly unmediated life. Increasingly, the rejection or abandonment of social media (and, in some cases, other digital technologies) also plays a significant role in the production and maintenance of fame, providing celebrities with further opportunities to enact and demonstrate authenticity. Within the contemporary context of digital overload and social media backlash – defined by widespread recognition of the filtered, edited, and staged nature of many social media posts, and by debates around the impacts of social media on mental health – these platforms are increasingly associated with the inauthentic and the toxic.[10] Growing numbers of celebrities are now capitalizing on this tide of anti-social media sentiment by taking temporary breaks from these platforms, strategically distancing their brand and values from this toxicity. For the stars who refused to open social accounts in the first place, this moment of digital disillusionment has served to reinforce their authenticity. Indeed, stars such as Keanu Reeves, Renée Zellweger, and Cillian Murphy appear wise, far-seeing, and steadfast in their values.

The reasons why celebrities opt out of social media or refuse to open accounts are many and varied but often relate to concerns with privacy, mental health, or productivity. Additionally, some celebrities quit social media to make political statements. For example, following Elon Musk's controversial takeover of Twitter in late 2022 – after which he rebranded the platform as 'X', fired 80 per cent of employees, allowed the mass reactivation of banned accounts, and drastically reduced content moderation protections – a raft of celebrities made grandiose statements protesting Musk as they left the platform, in what might be considered progressive political action or simple virtue signalling, depending on one's own standpoint. Those celebrities – whom we may term 'righteous refusers' – included Elton John, Trent Reznor, David Simon, Gigi Hadid, Whoopi Goldberg and Shonda Rhimes (Mier, 2022).

Some practices of analogue celebrity, such as the outright refusal of social media, may nostalgically hark back to a pre-digital golden age of celebrity where the lives of famous people were shrouded in mystery. Of course, attempts among the famous to conceal their private lives from the press often creates intrigue (and sometimes this is the strategic intention of the celebrity). Opting out of social media can therefore serve as a valuable PR tactic, generating mystery, buzz, and interest among the press around the celebrity's private life and their reasons for avoiding social media, while also generating compassion and respect from fans.

A focus on media stars who disengage with social media invites reflection on the subversive potential of digital disconnection practices among celebrities. When a high-profile star refuses to join these platforms or takes a temporary break from their online activity, and when this is accompanied by the celebrity critiquing data privacy issues or the addictive design logics of social media, it can open discussions around the negative potential of social media and big tech, and potentially encourage others to opt out too (though there is little evidence to suggest this). Temporary abstention from social media may initially appear as a critical stance against data-driven capitalism. However, while these temporary detoxes may momentarily deprive big tech companies of some valuable data, these breaks are rarely permanent and are often undertaken so that the celebrity can rest, re-energize, and ultimately return online, ready to re-engage their fans and followers, generating viral interest (and value) for social media platforms when they do so. The narratives and actions of celebrities who engage in momentary detoxes from social media often reproduce or reinforce neoliberal models of governmentality, whereby individuals are positioned as accountable and responsible for their (mis)use of digital technology within the self-regulation society.

This is not to suggest that celebrities who fully and permanently opt out of social media, like Keanu Reeves and Renée Zellweger, radically challenge contemporary configurations of capitalism. To some degree, their permanent rejection of social media enables these celebrities to opt out of the data-driven value generation that forms the economic basis of these platforms (though viral content about their lack of social media presence itself generates value for the platforms). Even if a celebrity opts out of social media, they remain very much discussed (and, as such, indirectly 'present') on these platforms and online more broadly. Indeed, as we have suggested, practices of digital refusal often perpetuate press interest in the celebrity, ultimately reinforcing the circuitry of attention economy capitalism that underpins wider celebrity culture (an economy which itself is intimately entangled with social media and digital devices). Celebrities themselves rarely critique their role in this economy. Furthermore, as the rejection of the digital becomes increasingly fashionable, commodified, and lucrative within the context of neoliberal consumer culture, celebrities that refuse to use social media (or engage in digital detoxes) now play an important role within a wider economy of digital disconnection. Spawning news articles and serving as role models for offline living, their digital disengagement is packaged into consumable content that satisfies a growing cultural interest in – and market for – the non-digital.

As such, technology refusal in these contexts cannot simply be understood as a counter-hegemonic practice. Digital refusal among the rich and famous may challenge dominant values around technology (by calling into question common assumptions that digital technology represents 'progress' and 'innovation', or by subverting norms that equate social status with owning the latest upgrade), but the practices discussed here also reflect the structural inequalities that shape both celebrity culture and digital culture, highlighting the new digital divide that has emerged between those who have the wealth and privilege to opt out of digital society and those who do not.

While it may be tempting to opt out of social media, this is not an option for the vast majority of people who rely on these platforms to find jobs, conduct work, maintain relations with family and friends, and engage with public services. Wealth and privilege enable analogue celebrities to log out and detach themselves from the realities of daily life in digital societies. Media theorists Simone Natale and Emiliano Treré (2020), among others (Karppi, 2011; Mejias, 2013; Rosamond, 2017) have argued that engagement (rather than disengagement) with digital technologies is a necessary precondition for developing adequate critiques of – and resistance to – the hegemony of the digital within contemporary societies. What is needed are strategies of resistance that are not based on withdrawal but on undermining or reworking the operating logics of social media platforms and the wider attention economy. Rather than incite change in digital society through critical engagement, the celebrity who disengages from the digital world retreats into their own luxury, private analogue idyll.

Notes

1 One wonders whether the campaign would have been more successful if, for every donation received, the celebrities had to spend longer offline.

2 At the opposite end of the spectrum to digital refusal, there is the chronically connected celebrity, represented by the likes of Snoop Dogg, Zendaya, and Khloé Kardashian, who release upwards of five hundred social media posts per month (Leskin, 2019). Beyond social media, some celebrities are reportedly very attached to their mobile phones. Back in 2008, in an interview with *Elle* magazine, Madonna revealed that she and Guy Ritchie (her husband at the time), 'were so busy that both of them slept with their Blackberry under their pillows' (Freeman, 2009: 164).

3 Perhaps the peak of celebrity engagement with social media was captured by the famous 2014 Oscars 'selfie' taken by Bradley Cooper and posted on Ellen DeGeneres' Twitter.

4 See, for example, the special issue of *Celebrity Studies* dedicated to Reeves, including an article on Reeves as a 'reluctant eco-celebrity' (Moffat and Kääpä, 2022).

5 For a number of celebrities, the manufacturing and maintenance of 'mystery' is perceived as a key facet of stardom. In a March 2024 interview, British actor Theo James discusses his lack of social media presence, similarly stating that 'I'm quite a private person … I like to keep a little bit of mystery' (Vij, 2024). Elsewhere, Chris Hemsworth has also commented that 'The mystery of who you are is what keeps people interested in wanting to see you on the screen' (Duboff, 2013).

6 An extreme, almost parodic, example of this comes from Ricky Gervais' social media feeds in which he posts purposely unflattering photographs of himself while having a bath.

7 For an analysis of celebrities criticizing the hollowness and superficiality of social media, see Ingleton and York (2019).

8 Some social media influencers do leave social media for good (and attract considerable attention in doing so). In 2015, the Australian Instagram influencer, Essena O'Neill, suddenly deleted many of her pictures and left the platform, stating that it was 'contrived perfection made to get attention' (Hunt, 2015; see also Jorge and Pedroni, 2021: 74–75). The act generated considerable attention among the international press and O'Neill was celebrated for raising awareness of issues related to body dysmorphia and social media addiction. It was quitting social media itself that enabled O'Neill to ascertain a new level of fame that transcended Instagram, for a while. Indeed, four years later, after sinking into relative obscurity, she returned to Instagram.

9 Sheeran is not the only celebrity to detox from social media only to return when strategically beneficial. Lorde uses social media in a similar way, releasing cryptic messages to build hype for her music after taking time offline (Paul, 2024).

10 As suggested, for example, by the rise of #NoFilter and 'Instagram vs reality' trends.

References

Abidin, C. (2018) *Internet Celebrity: Understanding Fame Online*. Emerald Publishing Ltd.

Abidin, C. (2017) '#familygoals: Family Influencers, Calibrated Amateurism, and Justifying Young Digital Labor', *Social Media + Society*, 3(2): 1–15.

Access Hollywood (2019) 'Britney Spears Slams Rumors That She Doesn't Post Her Own Social Media Content', *YouTube*, [online] 29 May, Available from: https://www.youtube.com/watch?v=AkY2xNqtkLk [Accessed 3 June 2023].

Bahou, O. (2021) '21 Celebrities Who Don't Use Social Media', *InStyle*, [online] 28 September, Available from: https://www.instyle.com/celebrity/celebrities-who-dont-use-social-media [Accessed 10 June 2023].

Banet-Weiser, S. (2012) *Authentic™: The Politics of Ambivalence in a Brand Culture*. New York University Press.

Bauerlein, M. (2011) *The Digital Divide: Arguments for and Against Facebook, Google, Texting, and the Age of Social Networking*. Penguin.

Baumer, E.P.S., Ames, M.G., Brubaker, J.R., Burrell, J., and Dourish, P. (2014) 'Refusing, Limiting, Departing: Why We Should Study Technology Non-use', *CHI EA '14: CHI '14 Extended Abstracts on Human Factors in Computing Systems*, pp 65–68.

Bendix, R. (1997) *In Search of Authenticity: The Formation of Folklore Studies*. University of Wisconsin Press.

Carty-Williams, C. (2021) 'Adele meets Queenie author Candice Carty-Williams to talk music, motherhood, dogs, divorce (babe, divorce) – and her record-breaking comeback', *The Face*, [online] 18 November, Available from: https://theface.com/music/adele-new-album-30-candice-carty-williams-interview-volume4-issue9-burberry-stormzy [Accessed 18 March 2022].

Cashmore, E. (2006) *Celebrity/Culture*. Routledge.

CBS News (2023) 'Margot Robbie on her Letter to Quentin Tarantino, First Paycheck and More', *YouTube*, [online] 29 December, Available from: https://www.youtube.com/watch?v=OmGG8dSXAIY [Accessed 3 April 2022].

Cheng, V.J. (2004) *Inauthentic: The Anxiety over Culture and Identity*. Rutgers University Press.

Chia, A., Jorge, A., and Karppi, T. (eds) (2021) *Reckoning with Social Media: Disconnection in the Age of the Techlash*. Rowman and Littlefield.

Cohen, E. (1988) 'Authenticity and Commoditization in Tourism', *Annals of Tourism Research*, 15: 371–386.

Crary, J. (2013) *24/7: Late Capitalism and the Ends of Sleep*. Verso.

Dare-Edwards, H.L. (2014) '"Shipping Bullshit": Twitter Rumours, Fan/Celebrity Interaction and Questions of Authenticity', *Celebrity Studies*, 5(4): 521–524.

Duboff, J. (2013) '"Why I'm Not On Twitter": Every Movie-Star Reason for Shunning Social Media', *Vanity Fair*, [online] 13 November, Available from: https://www.vanityfair.com/hollywood/2013/11/movie-stars-shunning-social-media-twitter [Accessed 3 April 2023].

Enli, G. (2015) *Mediated Authenticity: How Media Construct Reality*. Peter Lang.

Entertainment Tonight (2023) 'Ed Sheeran Returns to Social Media After "Turbulent" Personal Issues', *YouTube*, [online] 1 February, Available from: https://www.youtube.com/watch?v=1RrAVy3I9zE [Accessed 18 April 2023].

ET Live (2019) 'Why Britney "Feels Free" Posting on Instagram', *YouTube*, [online] 2 July, Available from: https://www.youtube.com/watch?v=fFwn4xbkpto [Accessed 28 March 2023].

Fast, K., and Syvertsen, T. (2024)' Post-digital Consumption: The Controversy Surrounding the Mobile Phone Box as a Means of Disconnection', in K. Albris, K. Fast, F. Karlsen, A. Kaun, S. Lomborg, and T. Syvertsen (eds) *The Digital Backlash and the Paradoxes of Disconnection*. Nordicom, pp 45–66.

Fish, A. (2017) 'Technology Retreats and the Politics of Social Media', *TripleC: Communication, Capitalism and Critique*, 15(1): 355–369.

Forde, E. (2015) 'Post-internet: Has the Bubble Burst on Pop's Social Media Strategy?', *The Guardian*, [online] 14 December, Available from: https://www.theguardian.com/music/2015/dec/14/post-internet-has-the-bubble-burst-on-pops-social-media-strategy [Accessed 18 March 2023].

Fowles, J. (1992) *Starstruck: Celebrity Performers and the American Public*. Smithsonian Books.

Freeman, J. (2009) *The Tyranny of E-mail: The Four-Thousand-Year Journey to Your Inbox*. Scribner.

Frowijn, L., Harbers, F., and Broersma, M. (2023) 'The Authenticity Gap: How Influencers Commodify Authenticity on Instagram', in M. Heřmanová, M. Skey, and T. Thurnell-Read (eds) *Cultures of Authenticity*. Emerald, pp 215–230.

Gaden, G. and Dumitrica, D. (2014) 'The "Real Deal": Strategic Authenticity, Politics and Social Media', *First Monday*, 20(1), Available from: https://firstmonday.org/ojs/index.php/fm/article/view/4985/4197 [Accessed 3 May 2023].

Gamson, J. (1994) *Claims to Fame: Celebrity in Contemporary America.* University of California Press.

Gozzi, L. (2024) 'Queen Margrethe: Will Abdication Cause a Ripple Effect?', *BBC News*, [online] 14 January, Available from: https://www.bbc.co.uk/news/world-europe-67944716 [Accessed 3 May 2023].

Glatt, Z. (2022) 'Precarity, Discrimination and (In)visibility: An Ethnography of "The Algorithm" in the Influencer Industry', in E. Costa, P.G. Lange, N. Haynes, and J. Sinanan (eds) *The Routledge Companion to Media Anthropology.* Routledge, pp 544–556.

Glatt, Z. (2023) 'The Intimacy Triple Bind: Structural Inequalities and Relational Labour in the Influencer Industry', *European Journal of Cultural Studies*, 27(3): 424–440.

Gurley, A. (2023) 'Ed Sheeran Says His Years of Social Media Absence Has to Do With "Turbulent" Things in His Personal Life', *BuzzFeed*, [online] 31 January, Available from: https://www.buzzfeed.com/alexgurley/ed-sheeran-social-media-return [Accessed 18 April 2023].

Gräve, J-F. (2017) 'Exploring the perception of influencers vs. traditional celebrities: Are social media stars a new type of endorser?' *Proceedings of the 8th international conference on Social Media and Society*, pp 1–5.

Hamad, H. (2018) 'Celebrity in the Contemporary Era', in A. Elliott (ed) *Routledge Handbook of Celebrity Studies.* Routledge, pp 44–57.

Hardey, M., and Atkinson, R. (2018) 'Disconnected: Non-Users of Information Communication Technologies', *Sociological Research Online*, 2(23): 553–571.

Helsper, E.J. and Reisdorf, B.C. (2013) 'A Quantitative Examination of Explanations for Reasons for Internet Nonuse', *Cyberpsychology, Behavior, and Social Networking*, 6(2): 94–99.

Heřmanová, M., Skey, M., and Thurnell-Read, T. (2023) *Cultures of Authenticity.* Emerald.

Hesselberth, P. (2018) 'Discourses on Disconnectivity and the Right to Disconnect', *New Media & Society*, 20(5): 1994–2010.

Horeck, T. (2022) '"Too Good for this World": Keanu Reeves, God of the Internet', *Celebrity Studies*, 13(2): 143–158.

Hunt, E. (2015) 'Essena O'Neill quits Instagram claiming social media "is not real life"', *The Guardian*, [online] 3 November, Available from: https://www.theguardian.com/media/2015/nov/03/instagram-star-essena-oneill-quits-2d-life-to-reveal-true-story-behind-images [Accessed 18 March 2023].

Ingleton, P. and York, L. (2019) 'From Clooney to Kardashian: Reluctant Celebrity and Social Media', *Celebrity Studies*, 10(3): 364–379.

Jorge, A., and Pedroni, M. (2021) '"Hey! I'm Back after a 24h #DigitalDetox!": Influencers Posing Disconnection', in A. Chia, A. Jorge, and T. Karppi (eds) *Reckoning with Social Media: Disconnection in the Age of the Techlash*. Rowman and Littlefield, pp 63–84.

Karppi, T. (2011) 'Digital Suicide and the Biopolitics of Leaving Facebook', *Transformations: Journal of Media and Culture*, 20: 1–18.

Kennedy, M. (2015) 'Ed Sheeran Quits Social Media – For Now', *The Guardian*, [online] 13 December, Available from: https://www.theguardian.com/music/2015/dec/13/ed-sheeran-quits-social-media-instagram-world-tour [Accessed 2 June 2024].

Kuntsman, A. and Miyake, E. (2019) 'The Paradox and Continuum of Digital Disengagement: Denaturalising Digital Sociality and Technological Connectivity', *Media, Culture & Society*, 41(6): 901–913.

Kuntsman, A., Martin, S., and Miyake, E. (eds) (2023) *Digital Disengagement: COVID-19, Digital Justice and the Politics of Refusal*. Bristol University Press.

Lasimone, A. (2021) 'Why Adele Had Her Instagram Password Taken Away', *Billboard*, [online] 5 December, Available from: https://www.billboard.com/music/pop/adele-instagram-password-live-1235006025/ [Accessed 14 April 2023].

Leskin, P. (2019) 'From the Kardashians to Snoop Dogg, Here Are the 25 Celebrities Most Addicted to Social Media', *Business Insider*, [online] 28 May, Available from: https://www.businessinsider.com/celebrities-who-use-instagram-twitter-most-kardashians-snoop-dogg-2019-5 [Accessed 16 August 2023].

Maares, P., Banjac, S., and Hanusch, F. (2021) 'The Labour of Visual Authenticity on Social Media: Exploring Producers' and Audiences' Perceptions on Instagram', *Poetics*, 84: 1–10.

MacCannell, D. (1973) 'Staged Authenticity: Arrangements of Social Space in Tourist Settings', *American Journal of Sociology*, 79(3): 589–603.

Marwick, A., and boyd, d. (2011) 'To See and Be Seen: Celebrity Practice on Twitter', *Convergence*, 17(2): 139–158.

Marwick, A. (2013) *Status Update: Celebrity, Publicity, and Branding in the Social Media Age*. Yale University Press.

McRae, S. (2017) 'Get Off My Internets: How Anti-Fans Deconstruct Lifestyle Bloggers' Authenticity Work', *Persona Studies*, 3(1): 13027.

Mejias, U.A. (2013) *Off the Network: Disrupting the Digital World*. University of Minnesota Press.

Mier, T. (2022) 'All the Celebrities Who've Quit Twitter Because of Elon Musk', *Rolling Stone*, [online] 19 December, Available from: https://www.rollingstone.com/culture/culture-lists/elon-musk-twitter-celebrities-quit-1234634670/elton-john-32-1234644683/ [Accessed 24 October 2025].

Moffit, K. and Kääpä, P. (2022) 'Taking the Green Pill? Keanu Reeves as "Reluctant Eco-Celebrity"', *Celebrity Studies*, 13(2): 200–213.

Morin, E. (1960) *The Stars: An Account of the Star-System in Motion Pictures*. Trans. R. Howard: Grove Press.

Muntean, N. and Petersen, A.H. (2009) 'Celebrity Twitter: Strategies of Intrusion and Disclosure in the Age of Technoculture', *Media/Culture Journal*, 12(5), Available from: https://journal.media-culture.org.au/index.php/mcjournal/article/view/194 [Accessed 3 April 2023].

Murphy, D. (2016) 'EXCLUSIVE: Renee Zellweger and Keanu Reeves Explain Why They Hate Social Media', *Entertainment Tonight*, [online] 21 October, Available from: https://www.etonline.com/news/201010_renee_zellweger_and_keanu_reeves_explain_why_they_hate_social_media_exclusive [Accessed 18 May 2023].

Natale, S. and Treré, E. (2020) 'Vinyl Won't Save Us: Reframing Disconnection as Engagement', *Media, Culture & Society*, 42(4): 626–634.

Oganesyan, N. (2021) 'Adele Said She Posted One Instagram Photo, and Her Team Responded by Taking Her Passwords "Away from Me Again"', *BuzzFeed*, [online] 5 December, Available from: https://www.buzzfeed.com/natalieoganesyan/adele-not-allowed-posting-social-media-passwords [Accessed 17 May 2023].

Outstanding Screenplays (2024) 'Cillian Murphy Teaches You How to Go Minimalist Black on Your iPhone', *Instagram*, [online] 24 March, Available from: https://www.instagram.com/reel/C44YH5KiSfr/?igsh=emltdHJsaWJ3MnQ%3D [Accessed 3 June 2024].

Paul, L. (2024) 'Lorde Teases Her Creative Path Forward with Cryptic Messaging and Hieroglyphics in Social Media Return', *Rolling Stone*, [online] 5 June, Available from: https://www.rollingstone.com/music/music-news/lorde-fourth-album-social-media-return-1235033457/ [Accessed 18 June 2024].

Peterson, R.A. (1999) *Creating Country Music: Fabricating Authenticity*. University of Chicago Press.

Pooley, J. (2010) 'The Consuming Self from Flappers to Facebook', in M. Aronczyk and D. Powers (eds) *Blowing Up the Brand: Critical Perspectives on Promotional Culture*. Peter Lang, pp 71–89.

Portwood-Stacer, L. (2013) 'Media Refusal and Conspicuous Non-Consumption: The Performative and Political Dimensions of Facebook Abstention', *New Media & Society*, 15(7): 1041–1057.

Press Association (2015) 'Stephen Fry Quits Instagram: "Hounded Off. Goodbye"', *The Guardian*, [online] 30 March, Available from: https://www.theguardian.com/media/2015/mar/30/stephen-fry-quits-instagram-twitter [Accessed 28 May 2023].

Quora (2019) 'Why Doesn't Keanu Reeves Use Social Media?', *Quora*, [online] 25 June, Available from: https://www.quora.com/Why-doesnt-Keanu-Reeves-use-social-media/log [Accessed 12 June 2023].

Rapkin, M. (2022) 'Renée Zellweger Lets Down Her Guard. (Sort Of.)', *Harper's Bazaar*, [online] 24 March, Available from: https://www.harpersbazaar.com/culture/features/a39435210/0001-0105-renee-zellweger-lets-down-her-guard-sort-of-april-2022/ [Accessed 3 April 2023].

Reade, J. (2020) 'Keeping it Raw on the 'Gram: Authenticity, Relatability and Digital Intimacy in Fitness Cultures on Instagram', *New Media & Society*, 23(3): 535–553.

Redmond, S. (ed) (2010) *The Star and Celebrity Confessional*. Routledge.

Reid, R. (2020) 'Adele Shouldn't Be Cancelled – But She Should Quit social media immediately', *The Telegraph*, [online] 2 September, Available from: https://www.telegraph.co.uk/music/artists/adele-shouldnt-cancelled-should-quit-social-media-immediately/ [Accessed 14 April 2024].

Riley, J. (2012a) 'Christopher Nolan Embraces Film Tech But Doesn't Own a Cell Phone', *Backstage*, [online] 15 August, Available from: https://www.backstage.com/magazine/article/christopher-nolan-embraces-film-tech-cell-phone-53210/ [Accessed 3 May 2023].

Riley, J. (2012b) '"Dark Knight Rises" Director Christopher Nolan's Shocking Admission: No Cell Phone, Email Address', *The Hollywood Reporter*, [online] 19 July, Available from: https://www.hollywoodreporter.com/movies/movie-news/dark-knight-rises-christopher-nolan-batman-352120/ [Accessed 3 May 2023].

Robinson, J. (2016) 'What Does It Mean? Ed Sheeran Shares His First Tweet in Exactly a Year and Fans Go into Meltdown Speculating What It Means', *The Sun*, [online] 13 December, Available from: https://www.thesun.co.uk/tvandshowbiz/2386167/ed-sheeran-shares-his-first-tweet-in-exactly-a-year-and-fans-go-into-meltdown-speculating-what-it-means/ [Accessed 12 April 2023].

Rojek, C. (2012) *Fame Attack: The Inflation of Celebrity and its Consequences*. Bloomsbury.

Rojek, C. (2015) *Presumed Intimacy: Para-social Relationships in Media, Society and Celebrity Culture*. Polity Press.

Rosamond, E. (2017) 'Reputation Addiction', in V. Bartlett and H. Bowden-Jones (eds) *Are We All Addicts Now? Digital Dependence*. Liverpool University Press, pp 97–103.

Satchell, C. and Dourish, P. (2009) 'Beyond the User: Use and Non-use in HCI', *OZCHI '09: Proceedings of the 21st Annual Conference of the Australian Computer-Human Interaction Special Interest Group: Design: Open 24/7*, pp 9–16.

Savage, M. (2017) 'This Is Why Ed Sheeran Quit Twitter', *BBC News*, [online] 4 July, Available from: https://www.bbc.com/news/entertainment-arts-40491956 [Accessed 12 April 2023].

Selwyn, N. (2006) 'Digital Division or Digital Decision? A Study of Non-Users and Low-Users of Computers', *Poetics*, 34 (4–5): 273–292.

Senft, T.M. (2008) *Camgirls: Celebrity and Community in the Age of Social Networks*. Peter Lang.

Seymour, R. (2019) The *Twittering Machine*. Indigo Press.

Smith, N. (2020) 'Christopher Nolan on His Viewing Habits, the "Hard Job" of Making Films and Why He Doesn't Email', *People*, [online] 12 December, Available from: https://people.com/movies/christopher-nolan-viewing-habits-making-movies-not-using-email/ [Accessed 16 April 2023].

Sprankles, J. and Gray, D. (2023) '29 Celebrities Who Reject Social Media & Why', *SheKnows*, [online] 5 April, Available from: https://www.sheknows.com/entertainment/slideshow/9721/celebrities-not-on-social-media/ [Accessed 23 April 2023].

Sutton, T. (2017) 'Disconnect to Reconnect: The Food/Technology Metaphor in Digital Detoxing', *First Monday*, 22(6), Available from: https://firstmonday.org/ojs/index.php/fm/article/view/7561 [Accessed 16 April 2023].

Sutton, T. (2020) 'Digital Harm and Addiction: An Anthropological View', *Anthropology Today*, 36(1): 17–22.

Syvertsen, T. (2020) *Digital Detox: The Politics of Disconnecting*. Emerald.

Syvertsen, T. and Enli, G. (2020) 'Digital Detox: Media Resistance and the Promise of Authenticity', *Convergence*, 26(5–6): 1269–1283.

teddysphotos (2015) 'Untitled', *Instagram*, [online] 13 December, Available from: https://www.instagram.com/p/BN8oFZEgleJ/ [Accessed 3 June 2023].

teddysphotos (2017) 'Hello 2017 …', *Instagram*, [online] 1 January, Available from: https://www.instagram.com/p/BOt_UFOl23o/ [Accessed 3 June 2023].

teddysphotos (2019) 'Brb x.', *Instagram*, [online] 24 December, Available from: https://www.instagram.com/p/B6dFOwZB6Jn/ [Accessed 3 June 2023].

teddysphotos (2022) 'Signing Off Now Until 2023. See You in the New Year. Much love, Ed x.', *Instagram*, [online] 4 November, Available from: https://www.instagram.com/p/CkiO70JI4qx/?hl=en-gb [Accessed 3 June 2023].

teddysphotos (2023a) 'F64 for SBTV', *Instagram*, [online] 19 January, Available from: https://www.instagram.com/p/Cnm_o6xKf5T/ [Accessed 3 June 2023].

teddysphotos (2023b) 'Back in the bits', *Instagram*, [online] 31 January, Available from: https://www.instagram.com/p/CoEs37fBDTo/ [Accessed 3 June 2023].

The Late Late Show with James Corden (2017) 'Ed Sheeran Carpool Karaoke', *YouTube*, [online] 7 June, Available from: https://www.youtube.com/watch?v=ckdsJ-LaCvM [Accessed 23 June 2023].

The Minimalists (2023) 'It's OK to Opt Out Completely', *TikTok*, [online] 2 September, Available from: https://www.tiktok.com/@theminimalists/video/7273842137513381162?_r=1&_t=8g48QcGrvH7 [Accessed 18 March 2023].

Thomas, S. (2014) 'Celebrity in the "Twitterverse": History, Authenticity and the Multiplicity of Stardom Situating the "Newness" of Twitter', *Celebrity Studies*, 5(3): 242–255.

Thompson, C. (2008) 'Brave New World of Digital Intimacy', *The New York Times*, [online] 5 September, Available from: https://www.nytimes.com/2008/09/07/magazine/07awareness-t.html?r=1 [Accessed 2 June 2023].

Trilling, L. (1972) *Sincerity and Authenticity*. Harvard University Press.

Turner, G. (2004) *Understanding Celebrity*. Sage.

Vij, A. (2024) 'The Gentlemen Star Theo James Gives Refreshing Reason Why He's Not on Social Media', *Lad Bible*, [online] 16 March, Available from: https://www.ladbible.com/entertainment/film/the-gentlemen-netflix-theo-james-social-media-410824-20240316 [Accessed 21 April 2024].

Watts, M. (2024) 'Woody Harrelson Says He Doesn't Own a Cell Phone to Not "Be Readily Available to Any Human Being at Any Time"', *People*, [online] 21 June, Available from: https://people.com/woody-harrelson-says-he-doesnt-own-a-cell-phone-8667212 [Accessed 2 July 2025].

Wootton, D. (2017) 'Sheeran twitter quitter Ed Sheeran forced to quit Twitter after revealing vile comments bring him down', *The Sun*, [online] 3 July, Available from: https://www.thesun.co.uk/tvandshowbiz/3939544/ed-sheeran-forced-to-quit-twitter-after-revealing-vile-comments-bring-him-down/ [Accessed 14 March 2023].

Wyatt, S. (2003) 'Non-Users Also Matter: The Construction of Users and Non-Users of the Internet', in N. Oudshoorn and T. Pinch (eds) *How Users Matter: The Co-construction of Users and Technology*. MIT Press, pp 67–79.

Wyatt S., Thomas, G., and Terranova, T. (2002) 'They Came, They Surfed, They Went Back to the Beach: Conceptualising Use and Non-use of the Internet', in S. Woolgar (ed) *Virtual Society? Technology, Cyberbole, Reality*. Oxford University Press, pp 23–40.

York, L. (2018) *Reluctant Celebrity: Affect and Privilege in Contemporary Stardom*. Palgrave Macmillan.

Zuboff, S. (2019) *The Age of Surveillance Capitalism: The Fight for a Human Future at the New Frontier of Power*. Profile.

7

Selling the Promise of Presence: How Companies Promote Digital Detox 'Products' in Denmark

Malene Hornstrup Jespersen, Annika Isfeldt, and Kristoffer Albris

Introduction

A growing body of research in the social sciences and media studies deals with the emerging trends of digital detoxing and disconnection practices in the context of pervasive digital devices and online 'fatigue' (Syvertsen, 2020; Chia et al, 2021). Research has been carried out on topics such as the social and ritualistic aspects of digital detox courses and 'camps' (Fish, 2017; Sutton, 2017), on the experiences of those who disconnect from various platforms and devices (Selwyn, 2006; Baumer et al, 2013; Portwood-Stacer, 2013; Jorge, 2019), and on the discourses around specific disconnection products (Syvertsen, Karlsen, and Bolling, 2019; Karppi, Chia, and Jorge, 2021) and in self-help books and similar texts (Syvertsen and Enli, 2020). In this chapter we aim to add to this growing interest in practices of digital detoxing and disconnection, taken to be a part of the wider political economy of attention in the digital age (Pedersen, Albris, and Seaver, 2021), in which our attention (and lack of attention span) is increasingly perceived as a matter of concern (Williams, 2018). We do this by homing in on an aspect that has received less focused attention, namely the companies that offer digital detox and disconnection services and products. While the large majority of the growing literature on digital disconnection focuses on the end-users or clients, less attention is paid to the businesses that are part of the digital detox trend and the entrepreneurs that have established these companies. This chapter seeks to fill that gap.

The analysis is based on in-depth interviews with five of the most prominent companies in Denmark that offer digital detoxing as a 'product'. The scouting process was conducted online, sampling only companies explicitly using the phrase 'digital detox' in the sales material describing their products and services. We combine the interviews with an analysis of the companies' websites, including the way they discursively advertise digital detox as a product and the aesthetics of the images they use to attract and retain the attention of potential clients towards these products.

We examine how these companies seek to latch on to a particular kind of disconnection Zeitgeist (Albris et al, 2024) related to themes such as mindfulness, work-life balance, inner formative journeys, and a return to nature, all of which are couched within an overall concern with the increasing digital consumption of present-day society. Central to such a concern is an explicit discursive and practice-oriented engagement with different notions of what constitutes the analogue as an anti-thesis to the digital. Importantly, several of the digital detox and disconnection 'products' that the companies promote, are in one way or the other presuming the mental and physical health superiority of all things analogue, meaning non-digital. Importantly, a primary selling point for the companies is their promise to clients that they can attain a more authentic presence in the world without digital devices, and that this authentic presence is achieved via analogue means.

While the five companies we examine do not in themselves constitute a fully fleshed market or an economy as such, they point towards both existing and future trends in which 'the digital' is increasingly being framed by professionals in the mindfulness and meditation industries as a symbolic negation of what the good life entails. Combatting the digital thus becomes something one can commodify and sell. And, as we highlight in this chapter, one of the notions that lie at the core of this negative assessment of digital devices is that they remove people's ability to be present in the world. Hence, the companies we focus on in this chapter share a common understanding of selling the promise of presence to their clients and customers.

The chapter begins by explicating the term *digital detox* according to its etic and emic meanings, and continues by describing the businesses that constitute our research sample and how they view the market for digital detox in Denmark. The remaining part of the chapter delves more deeply into *presence* as a recurring theme promoted by the company owners as a key element of digital detoxing. We identify three types of presence that digital detox companies refer to in their course material and underlying philosophies: 1) presence with oneself, 2) presence with others, and 3) presence with the world around us (or the environment(s) we inhabit). Finally, we conclude by considering how these ideas of presence are interconnected, and how the companies we have looked at in the Danish context relate to the wider and emerging field of disconnection studies.

Digital detoxing in Denmark and beyond

In the first section, we begin with a description of the profiles and aims of the companies, focusing on the variety of methods they employ to forge connections between the aims of digital disconnection and other moral and political concerns. As a preface to that, we briefly discuss how digital detox as a phenomenon and idea is emically conceptualized by the companies.

Digital detoxing as a societal trend

The term *digital detox* has emerged in the last few decades as a way to describe people taking periodic breaks from their everyday use of digital devices and platforms. Digital detoxing is part of a broader spectrum of terms denoting an awareness of the ills of digital devices, such as going 'off-the-grid', digital self-control (Lyngs et al, 2020), and digital disconnection more broadly. However, digital detox points more specifically to a *temporary* pausing of people's digital consumption in order for them to become 'detoxified' (Syvertsen, 2020). Thus, in contrast to the aim of leading an entirely non-digital life, the point of digital detoxing is to remove oneself from the digital only temporarily, with the intention of returning. This temporary aspect means that what is conceptualized as digital detox is often practiced as part of social settings such as retreats in nature (Fish, 2017; Sutton, 2017), sessions that are part of longer training programmes, such as mindfulness training (Cook, 2016) or ongoing consultations with self-proclaimed experts in various versions of cognitive training. Following the popularization of the term, digital detox is now also commonly used to denote many digital self-control practices, including limiting screen time and use of social media for oneself or one's children, and not just the periodic suspension of the use of these (Syvertsen and Enli, 2020).

The term 'digital detox' is commonly used in the Danish language as an anglification, and there is no particular Danish understanding of the term that diverts from that found in the UK or the US contexts. The trend of going on digital detox retreats or signing up for longer training programmes, however, seems to be less prevalent in Denmark than in countries such as the United States if one merely looks at the small number of programmes and courses available to potential detoxing clients. Yet there are currently no studies that have investigated this at a larger scale, apart from individual studies in certain countries such as Norway (Syvertsen, Karlsen, and Bolling, 2019).

This study is one of the first of its kind with respect to empirically investigating digital detox practices and discourses in Denmark (see also Bagger and Lomborg, 2021), albeit from the perspective of businesses offering digital detox products. As a result, we are here less interested in how one defines digital detox as an etic, general term, focusing instead on how selected companies in Denmark understand digital detox emically. In other words, we are interested in understanding what the term means for those offering it to clients, and how it attaches itself to other terms, concepts, and ideas, especially that of presence and attention.

For one of the companies we have looked at (Venture North), digital detox is defined both as a kind of antithesis to online life, while also being a matter of 'recharging'. As they write on their website: 'The alternative to the online-life is called digital detox. Here, you log off completely and go into nature to recharge.' An owner of another company (The Presence Movement) defines digital detoxing by focusing on the toxic semantics undergirding the term, which to her demands a process of cleansing oneself and excorcizing the digital demons from within:

> Detoxification. It means, again, that we need to detoxify ourselves. The things that are toxic to us, … the things that steal our good energy, that steal our time, that which doesn't benefit us, we need to get rid of that. … In this case, the focus is just on the digital. (Sophie, The Presence Movement)

As a third example, the owner of the company The Concentration Institute describes in an interview digital detoxing both as a preemptive measure to deal with stress, anxiety, and depression as well as a tool to be used if one has experienced any negative psychological states:

> I think there are a lot of things in digital detoxing to me, I mean that it both affects our mental health, but actually also something physical. And to me there's no doubt that it is both part of the preventive measures both toward stress and, actually, also anxiety and depression, and then also some of the tools needed to heal. (Marie, The Concentration Institute)

The same company owner subsequently adds that digital detoxing is also about controlling one's life and about presence: 'To me it means going a little more analogue, that you can focus your attention more on being in the moment' (Marie, The Concentration Institute).

As these quotes from the interviews and from the websites express, digital detox is a polysemic term that can be combined with a range of other terms and concepts, depending on the context of what a particular company aims to offer clients. As an introduction to the rest of this chapter, the previous quotes concerning the meaning of digital detox highlight that the companies we are investigating do not share a narrow common understanding of the aims and goals of digital detoxing, although there are some clear overlaps.

Introducing the companies

While many Danish companies offer services that indirectly include a distancing from digital device use (as for instance in mindfulness or meditation programmes), the five companies in our study were the only ones we could find that explicitly sell courses, programs, and trainings as digital detox products on the Danish market. The companies are located in different parts of the country, were not aware of each other, and were not able to introduce us to further businesses in the field, inhibiting a snowball sampling strategy. The fact that there are so few companies also testifies to the fact that the distinct trend of digital detox in a small country like Denmark is marginal.

The five companies have different backgrounds and objectives. They fall into two main categories: one consisting of companies with a base in the Eastern-inspired cosmologies of yoga, mindfulness, and nature therapy, often focusing on mental health and wellbeing; and another consisting of companies with a focus on outdoor living, crafting, and adventure, whose aim is to help their clients get 'back to basics' and discover a simpler way of life by surviving in nature without the help of digital devices.

Representing the first category in our sample are The Concentration Institute and Center for the Senses.[1] Marie owns The Concentration Institute, which focuses on helping individuals to deal with stress, anxiety, and depression. Their digital detox services are part of this agenda by aiming specifically to prevent stress. This is done through a variety of methods, spanning retreats, lectures, courses, and individual coaching. Retreat days often include research-based theory about how digital device usage affects the brain, followed by small exercises. Longer courses can include tracking of time spent online and exercises aiming to deal with cravings, centring around meditation, mindfulness, and nature connection. The company caters to both businesses and individuals over 30 years of age – presumably because this demographic age group has the resources to invest in retreats. Among these individuals are people explicitly seeking help to deal with what they see as digital addictions. The Concentration Institute is also a subcontractor to municipalities, delivering stress management courses, in which digital detox is included as one theme among many.

David owns Center for the Senses, a company that focuses on a self-developed body therapy and which offers a range of retreats and treatments, including cleansing and dietary guidance, meditation, yoga, and holistic spiritual retreats. Recently Center for the Senses launched a retreat focusing specifically on digital detox for men. Here, a group of men spend three days in the Jordanian desert away from any phone signal, receiving body therapy, doing yoga, and meditating. The trip also included two touristic days at the Dead Sea after the detox. The clients for this course were almost all wealthy businessmen between the ages of 35 and 61. Many were not used to practicing yoga or meditating; approximately half of Center for the Senses's digital detox clients had never tried yoga before. David's goal is to remove digital media to allow for the facilitation of self-reflection and self-actualization, which he believes can lead to better lives. By targeting men specifically, the hope is to attract clients that would perhaps be disincentivized to go on digital detox retreats in other contexts. By also targeting groups of men who already know one another, the aim was furthermore to tap into a trend of male self-care that are also found in other contexts, such as male adventure camps.

The businesses owned by Peter (Wilderness Survival Denmark) and Thomas (Venture North) represent the second category. Both offer wilderness survival courses and teambuilding exercises, where phones are locked away upon departure. The focus of these companies is to remind their clients of their most basic needs and nature survival skills. Thomas' Venture North also offers a range of adventure trips, where the focus is on out-of-the-ordinary outdoor activities. The main target groups for Wilderness Survival Denmark's digital detox services are companies seeking team building exercises in nature. Wilderness Survival Denmark have run courses for a wide variety of companies, but often the companies are Copenhagen-based and have a busy working style. Team-building exercises for companies consist of initially locking phones and computers away, and then giving the team a range of nature survival tasks, such as finding firewood and preparing a meal in the open, at the end of which they will win back the key to access their digital devices. Wilderness Survival Denmark also serves individual clients, mainly in its nature-survival related courses, which are digital-free.

Sophie owns the fifth and final business, The Presence Movement. This is the newest of the five and differs from the others in that it is the only company offering digital detox services as its primary product and in that it takes its point of departure in Sophie's background as a schoolteacher with a specialization in wellbeing and behaviour. Sophie's main target group is families with older children and teenagers, where the parents feel their children are 'missing out on life' due to excessive screen time, and who have had difficulties talking with their kids about this. She has also given lectures to groups of secondary school teachers. The services center around six-week-long online courses, consultations, and workshops. The courses are composed of 43 short videos accompanied by workbooks. Her methods also include dialogue-starter games centering around online life, tracking online habits, and helping her clients to organize their time spent online. Sophie's main mission is to help her clients take control of their use of digital devices, and to integrate the opportunities granted by online connectivity with her view of a healthy lifestyle.

Moving from this initial presentation of the five companies that form the sample for our analysis, in the following we try to provide a short introduction to what constitutes the digital detox market in the words of the company owners.

The digital detox market

The markets for digital detox products have been growing across the Western world in recent years, especially in global centers for technology such as Silicon Valley in the United States (Fish, 2017; Sutton, 2020). Denmark is one of the countries in the world with the highest level of public and private digitalization (UFM, 2020); however, some feel the pervasiveness of digital devices has become overwhelming (Jespersen, Albris, and Willadsen, 2025). A 2018 survey conducted by Deloitte showed that 35% of Nordics feel they use their smartphones too much, while for those in the age group 18–24, this is true for 54% (Malmlund et al, 2018).

Despite this prevalent public concern, the interviewed company owners experience the demand in Denmark for products explicitly advertising digital detoxing as small. Marie comments that 'there just hasn't been a demand for it as a stand-alone product'. Echoing this point, Peter states: 'It's a funny line of business this, because you can't just sit down and say, 'now I only want to make digital detox courses.' Denmark is just not big enough for that, so you have to spread out a little' (Peter, Wilderness Survival Denmark).

The company owners have different theories for why the demand is low. Marie thinks the phrase 'digital detox' might not seem appealing, as it could have a negative connotation, like smoking cessation, whereby some potential clients might be uninterested to be labelled as being addicted to their smartphones and other digital devices. Within the corporate sector, she theorizes that businesses are in an ambivalent position, wanting on one hand to create a healthy environment for their employees, but, on the other hand, not wanting them to stop reading emails in the evening.

As seen in the earlier quote, Peter attributes the lack of demand to the relatively small size of the country. However, other fields that presuppose digital disconnection, such as mindfulness courses, have in the past decades experienced a rise in demand in Denmark, high enough to sustain many successful companies (Borup and Jensen, 2015). Peter, Marie, and David have all noticed this trend. Peter sees the demand for nature therapy as 'booming' and identified this as part of a flourishing market for 'getting people away from this stressed everyday life'. Marie has noticed a similar interest, especially with the advent of the COVID-19 crisis:

> I think what's been most popular the past year, especially in 2020, because of corona, has been this booming interest in getting into nature in new ways, and really … maybe also finding this slowness, where you just stroll around and give yourself time to wake the senses. And that is of course indirectly digital detox, because when you wake the senses in that way in nature, you are offline. (Marie, The Concentration Institute)

Prior to the pandemic, yoga and meditation retreats had also been selling increasingly well for The Concentration Institute. Marie sees these trends as an expression of a generalized wish for inner formative journeys.

The company owners all see digital detoxing as a naturally integrated solution to other commonly problematized facets of modern life. All companies with the exception of The Presence Movement only offer few outright and distinct digital detox courses, but they all used analogue techniques in many of their other services. For instance, Marie integrates disconnection methods in her sleep hygiene courses, for instance by removing phones from bedrooms, and David and Peter see it as natural that their body therapy and nature crafting courses, respectively, were digital-free zones. Sophie from The Presence Movement is the exception in our sample as she is the only one selling stand-alone digital detox services as her primary product.

Our informants' experiences show that in the Danish market, digital detoxing is seldom the main selling point. Instead, digital breaks are packaged as part of other, more popular, agendas, such as stress reduction, nature connection, and facilitation of inner reflection. These themes are also apparent in the way the companies present themselves online, using catchwords such as 'authenticity', 'simplicity', and 'tranquility', and often appealing to potential clients with beautiful images of nature and of people in nature engaged in contemplation, meditation, or crafting activities. Going offline is integrated into – or a prerequisite – to these agendas but does not carry the sale.

Three of our informants feel that the problem of excessive connectivity goes unrecognized in large parts of society. According to Peter, many people do not realize that what they are 'truly looking for' when seeking trips into nature is the absence of the digital. Sophie was new to the market at the time of the interview, but her hope is to get through to the many Danish families that, according to her, do not recognize their need to (re)gain control over their use of digital devices. Despite this, she believes that the market is growing. David is, likewise, certain that the market will expand in the future, as he believes constant connectivity will make (frequent) digital breaks necessary for the modern human way of life: 'We have many different retreats, but this is the only one we call a digital detox, so ... there'll definitely be more of those, like that type of retreat, going forward, because it'll become a necessity for people to pull the plug, right?' (David, Center for the Senses).

In the previous sections we have introduced the companies and their different views of digital detox as a phenomenon. In the next sections we move on to investigate the deeper question of what undergirds their conceptions of digital detox as something they can offer clients as products. We do this by highlighting how different ideas about presence constitute a crosscutting theme for the companies.

Presence as the aim of digital detoxing

Presence is a central part of the online sales material for all five companies. The word is incorporated into the company name The Presence Movement, whose front website page also states 'Digital detox: less screen – more presence'. The Concentration Institute advertise their services by writing 'Take a break from deadlines, phones, tons of emails, everyday demands, and racing thoughts and become present in the moment!' 'Presence and stillness in nature' is part of the listed content in Venture North's two-day digital detox course. Additionally, both David and Marie explicitly state that presence in the moment is central to what they hope their clients will gain. When asked what attracts people to digital detox courses, David answered: 'I think, bottom-line, presence. Being more present and not being addicted to phones. I think it can be put that simply.'

The companies have different ideas of what is disrupting our ability to be authentically present in the world. Yet a common thread is the identification of being constantly (over-)connected. All five company owners have strong opinions about digital media and devices as being alienating to people, albeit they frame this in different ways. Some believe that digital devices hinder our ability to listen to our 'inner voice', provoke a mental distancing from others, or cause a loss of control over how we spend our time. The main tool to ameliorate the adverse effects of this alienation is *presence*, understood by our informants to be a meaningful, focused attention towards anything non-digital in both a spatial and temporal sense. We identified three types of presence in the way the company owners invoked the term: the ability to be present with ourselves (inner presence), to be present with others (social presence), and to be present in the world around us, or within the environment(s) we inhabit. In the following section we expand on the values and goals expressed by the companies when promoting digital detoxing through this central, tripartite idea of presence.

Inner presence: from mindlessness to mindfulness

Inner presence is an umbrella term for the ideas and practices expressed by the five company owners that target self-awareness and being in control. To be present with oneself is to be aware of our own basic needs and the true necessities in life, to be aware and in control of what we attend to, and a path toward being able to self-reflect and self-actualize. As digital media are viewed as a central hindrance to these goals, disconnection becomes a form of self-care.

Both Wilderness Survival Denmark and The Presence Movement had a focus on fundamental needs in their work. In Peter's view, the modern person has forgotten what their true needs are, and mistakenly sees digital devices as necessities, while he thought that in many cases, devices are more stressful than helpful. His main goal was to help people out of their busy everyday lives and bring them 'back' to a simpler way of living, where anything superfluous is removed. Removing digital media allowed for true relaxation and presence with oneself, he believed.

Sophie too fears that an excessive use of digital devices was causing the modern person to neglect the fundamental needs we have as humans, likening digital media to unhealthy food:

> If our digital lives obstruct us from getting our real needs fulfilled, health needs … It's kind of like, well nothing's going to happen if you eat pizza occasionally. But if you eat so much pizza that you don't get any broccoli and that kind of thing, you know, you have to cut down on the pizza. (Sophie, The Presence Movement)

The needs that might be obstructed by digital media were both physiological, including exercise and sleep, and mental, including presence with others, gaining new knowledge, and spirituality. In Sophie's view, the digital self-alienation starts early, as digital media in contemporary society from early childhood is linked to emotional dependence:

> A lot of parents have raised their children from when they were very small with an iPad as something to give when the kid is tired, when the kid needs to be comforted, when the kid in some way or the other needs to be regulated emotionally. And that means, it simply becomes such a strong symbol, actually, of emotional regulation, even though it sometimes does the exact opposite. … the reason that it can get such a high status in kids' brains is exactly because it takes over the control of the kids' brains, so we simply must help them regain that control. (Sophie, The Presence Movement)

Sophie draws on neuroscientific language in her critique of the use of digital devices for children, worrying that digital tools cause children to lose control over their own minds and affect their abilities to absorb knowledge, make good choices, be creative, be forward-thinking, and take initiative. This mirrors Sutton's (2020: 22) analysis of detoxers' claim of an unnatural modern world in which the human mind no longer evolves. Within this framework, it is the duty of adults to help their children – and themselves – be in control of digital tools, so the tools don't control us.

The loss of conscious control over our minds is not believed to be confined to children only. A main goal for both Sophie and Marie is to help their clients become aware of their use of digital media, and oppose unconscious habits, 'mindless zapping', and 'autopilot': 'I at least personally would like to create an awareness that we can use these as good tools, but we need to be careful not just to go on autopilot and then just mindlessly sit and zap … going from one app to the next' (Marie, The Concentration Institute).

Most of the company owners highlighted that digital detoxing is an individual activity and responsibility, which one might see as being part of general societal imaginary of the responsibility the individual person to care for and help oneself. The aim is to get their clients to recognize their own unconscious habits and unhealthy practices. Underpinning this aim is the idea that awareness and control of one's own attention will ideally lead to the ability to be present in the moment and a higher quality of life: 'So, my hope is really that we actually get a higher quality of life by getting a greater awareness about how we use it, our digitalization' (Marie, The Concentration Institute).

In The Presence Movement's and The Concentration Institute' digital detox courses, awareness was created through tracking and mapping clients' digital habits. Awareness about digital habits was seen to open space for presence, but presence was also used as a tool to create awareness. As part of Marie's sleep hygiene courses, she asks participants to disconnect for an hour before bed, and spend that time with themselves, doing a puzzle or something similar. She reports that this made her clients aware of their unconscious habits, that is, how they previously had spent time scrolling through cat videos or Instagram, and that one of their main takeaways was a feeling of 'having been gifted an extra hour' and 'enjoying this me-time'.

Of the five company owners, David has the most explicit focus on presence with our inner selves. His retreats focus on 'cultivating healthy egoism' and he sees disconnection as a way to 'prioritize oneself'. He believes that too much screen time inhibits self-reflection. He aims to create a physical framework that can help his clients connect with themselves and have an embodied experience of listening to themselves:

> As I usually say, a retreat with us is bottom line about opening your heart and getting better at listening to yourself. And you can't do that if you're looking into a screen all the time. You can tell that to the head, but the head will not understand until you have experienced it. (David, Center for the Senses)

Pushing participants towards self-actualization is the ultimate goal of the treatments David sells. For him, digital disconnection is a necessary, albeit preliminary step. Once the constant opportunity to go online is gone, methods such as yoga, body therapy, and social gatherings facilitate the internal and external presence participants seek. David's hope seems to be that being present with oneself, and the accompanying self-reflection, optimally leads to self-actualization and a better life for his clients.

Social presence: disconnection as reconnection

It is a common claim that the presence of digital devices damages social interactions. In fact, it has become so common that the term *phubbing* is now widely used to denote 'the act of snubbing someone in a social setting by looking at your phone instead of paying attention' (Merriam-Webster, nd). The suppliers of digital detox services ascribe to the idea that digital devices are damaging relationships and social interactions by being a disrupting and distracting interference in face-to-face situations. In the interview, David describes how Center for the Senses's desert-based detox retreat had been a social experience for the whole group of participants, but when the Wi-Fi was turned on in the bus on the way back, all heads dipped into their screens, and there was no conversation on the three-hour bus ride. Setting aside the possible social connections happening online or that phone-use may be a method of taking a social break and replenishing attentional resources (Coker, 2013), to David, the lack of conversation clearly shows that the opportunity to be connected is damaging to social presence. Embedded within the view of digital media as damaging to social situations is, then, an underlying position that people present physically should be prioritized over people present via digital connection. Marie describes a situation in which a salesclerk picked up the phone to service a client on the line, rather than attending to the client physically present (herself). This prioritization irritated her, as she believed the face-to-face interaction was more important.

The main issue, as identified by our informants, is that smartphones are used to create a physical and mental barrier that people hide behind to escape interacting with others. Marie believes digital devices can bring us together, but also make us withdraw into ourselves. To David, the smartphone is also a social barrier by being a marker of social status and a container for our 'ego':

> Try bringing twenty businessmen, they can often have a big ego. And just the fact that the phone and the suit is removed, some of the ego disappears. Then it's really just two Mr. Hansen talking, and then there's nothing about 'I earned this and this much money' and 'my company' and that kind of thing, those talks were not there at all. You couldn't feel who was rich and who was poor, and who was fat and thin, I mean, everyone was equal. And that was great. (David, Center for the Senses)

Stripping people of their phones is for David a central part of facilitating social connections that put people on an equal footing. However, as the majority of David's clients are wealthy, white businessmen, we might question whether this experience would in fact generalize to more unequal social relations. Nevertheless, this is at the heart of his business model.

One of Sophie's main concerns are the barriers created by digital devices between parents and children. She recalls an example from her teaching days when she had watched a movie with her fourth-grade class. The children had been surprised that she was really watching the movie, remarking that their parents were always on their phones when they watched movies at home. To Sophie, this showed that smartphones create a social distancing within families. Similarly, Sophie felt digital devices were creating a barrier between parents and their infants, as smartphone use in her experience has become prevalent when taking infants on stroller walks: 'It hurts my heart so much, you know, because I know there's a little child in there looking up and hoping to be seen by their mom and dad' (Sophie, The Presence Movement).

When considering older children, the conflict was flipped: according to Sophie, disagreements about the amount of screen time allowed to the children, was the source of conflict in the home. Yet both the barriers created by parents' use of smartphones and the conflicts caused by disagreements over screen time were damaging to social presence within the family.

David and Peter both emphasize that conversations on their courses and retreats are 'deeper' and 'better' when digital devices are not present: 'There were talks that were so much deeper than they otherwise would have been, because you didn't hide behind – you couldn't say, well I don't want to talk now, I'll just pull out my phone' (David, Center for the Senses).

The digital detox courses run by Wilderness Survival Denmark and Center for the Senses are actively aimed at creating spaces that can foster deep conversations and face-to-face connections, for instance by facilitating team building exercises, activities around campfires, and quiet togetherness looking at stars. Disconnection from devices is thus seen as a re-connection with others, mirroring previous studies on digital detoxing (Sutton, 2017). Being disconnected is seen as being 'liberating' and creates, in David's words, 'a lot more presence'.

The COVID-19 pandemic highlighted some paradoxical aspects inherent in this understanding of social presence, as physical participation in courses and programs were limited during lockdown, if not impossible. One case illustrated this particularly well: the Concentration Institute attempted to adapt to the COVID-19 lockdown by creating online video-call versions of nature-connection exercises. Marie had identified a need to alleviate loneliness through screens, stating that in this case, 'the screen actually … has something to offer'. Each client would call in via Zoom from a nature location of their choice, and Marie guided them in nature-connection exercises, creating a feeling of 'unity' among them. As Marie phrases it, during these online exercise sessions, clients were 'digitally connected, but present in the moment'. The situation of the pandemic flipped the idea of disconnection as reconnection on its head; suddenly, online connectivity meant social presence.

Presence with the world around us: connecting with the environment

The third type of presence is a presence with one's physical and temporal environment. Digital media disrupts our very sense of time and space, our informants in various ways claim. The disruption is grounded in the development of unconscious digital habits and that we lose touch with our surroundings and how we spend our time.

The company owners speak of our time as 'stolen' and 'wasted' by digital media. This wording carries with it an underlying judgment of how time is well spent and assumes that the individual could be doing something more valuable with his or her time. Both Marie and Sophie relate this loss of control over our time directly to a decrease in life quality. Marie describes how her clients were 'shocked' when they discovered how much time they spend online, and remarked that when online, they quickly lose track of time:

> It's this addiction that a great many people are not aware of, and the brain likes making these loops. Before we know where we are, then … after twenty minutes, we think maybe five minutes passed. Then when we check the different apps on the phone or iPad for instance, well then there's new notifications, so in that way it steals our time. (Marie, The Concentration Institute)

A balanced life requires that we make controlled, intentional choices about how we spend our time, according to our informants. What we need to get rid of, Sophie believes, is 'digital snacking', that is, unplanned, and often subconscious, use of our devices. Sophie saw the vast possibilities of online connection as a dangerous opportunity to escape from 'reality' and 'the world', understood to be our immediate environment. Instead, we spend our time in 'Lala-land':

> The danger is two things, I'd say. One thing is that there is no further than the small press of a button, then you're in Tivoli-land instead, then it's Lala-land, right, where your focus is stolen by all kinds of things and then often there are these notifications that tell you, oh now something or the other came up, something on Snapchat or something like that, right, then it's again this divided attention. (Sophie, The Presence Movement)

This constant possibility of escapism is in direct opposition to true temporal and physical presence in Sophie's view. Digital media alienate us not only from ourselves and each other, but from our very sense of temporal and physical surroundings.

A second problem framed by our informants is that of temporal overload; essentially that modern life is busy and stressful, and digital media are a major contributor to this. Syvertsen and Enli (2020) show that in most respects, the modern (Western) person has more leisure time than previously in history, and that there are many historical examples of worry over overload. Both mass-printing of books and satellite television prompted concern about informational overload when they became widespread. The main difference today is that online connectivity has become available almost everywhere and all the time (Syvertsen and Enli, 2020). According to David, the constant possibility for online connection has become naturalized to a degree that we no longer notice its pervasiveness: 'We don't notice, if you ask me, us in the Western world. Like, it's just something that ... we just have to be available all the time, right, then there's Apple Watch and ... there's no break' (David, Center for the Senses).

The lack of 'breaks' is what creates the experienced temporal overload. Our informants felt the constant connectivity contributed significantly to modern problems of stress, and it was central to what they believed their clients were seeking escape from.

The digital detox businesses had different exercises aiming to help their clients relax by regaining their sense of time and connection with their physical environments. Many of the exercises were centered around what David called 'empty' time. For instance, this manifested as leaving time unplanned in digital detox retreats:

> Otherwise, there's just space for absorption. Like, there are no plans. There were already two who Thursday morning said, 'well can't we drive to – we'd like to see the Dead Sea, it's only four hours each way, and we can do that in a day', and then I said, 'hello boys, why did you book?' I mean ... and then they said themselves that oh yeah they saw what I meant. (David, Center for the Senses)

Assisting his clients in accepting unplanned, 'empty' time, is important to David. This is also a main goal for Wilderness Survival Denmark: helping their clients to turn towards a simple lifestyle by keeping a focus on fulfilling basic needs, that is, food, water, and sleep, and the number of activities to a minimum.

There were also methods directly aimed at regaining a sense of passing time when unoccupied by other things. For instance, Marie often starts workshops with two minutes of meditation, 'to demonstrate how much happens in our heads when we just sit for two minutes'. Wilderness Survival Denmark use an exercise they called 'out-sitting', wherein participants try to regain an inner sense of time they may have lost in their digitally mediated lives and thereby find a deeper sense of relaxation:

> In all simplicity, the idea is that you find a nice place in the forest and sit there, without watch or phone or anything, and then we say, 'try to

> come back in an hour' and people are like, 'I can't, because I don't have my watch with me'. And then they have to try to get a feeling of how long an hour actually is when you're just sitting there. It's really quite fun, and there's actually, I mean, there's a big difference between how long an hour is for people, because some come back pretty precisely, within five-ten minutes or something, and others we've had to go get two-three hours later, because they're still sitting out there. (Peter, Wilderness Survival Denmark)

Nicola Johnson (2020) argues that smartphones enable us to enter a *temporal autonomous space* through which the individual has the agency to escape undesired situations and render their time more efficient – basically, spending time on their phones instead of attending to the space they are physically present in. In this way, Johnson theorizes smartphone use as a way for the individual to gain a feeling of control over a situation that they would not otherwise necessarily have control over. Rather than seeing smartphone use as choosing to enter a *temporal autonomous space*, our informants seem to view this space as something that grabs and withholds our attention, and choice as something that needs to be developed through awareness and the pruning of attentional self-control. Exercises like 'out-sitting' are meant to relocate the individual within a temporal presence – and a physical presence, in that there is nothing to do but be aware of the physical environment. Most importantly to Peter, it is meant to allow the individual to truly relax, free of the constant awareness of time and free of the possibility of digital temporal escape.

Conclusion

By way of conclusion, we discuss how the companies we have looked at in the Danish context relate to the wider and emerging field of disconnection studies. As stated before, most studies on digital disconnection focus on the customers of detox retreats, products, or services. While some studies, for example the work of Sutton (2017) and Fish (2017), do discuss the contexts of such detox retreats and the design ideas behind them, little has been written about the people who create companies that offer courses and retreats on disconnection.

This chapter contributes to current research on digital detoxing and disconnection from the perspective of the people who design courses, programs, and retreats, including the business models they pursue to make an income from their companies and the values they subscribe to in doing so. By focusing on companies offering digital detox services, the chapter engages directly with the ways in which ideas tied to analogue and non-digital living become packaged and sold as a product. As we show in the chapter, selling digital detox cannot be done without making reference to other trends and values, such as a return to nature or work-life balance. With widespread digitalization, a new frontier for self-optimization has opened up. Digital detoxing can be seen as part of a broader self-improvement culture, where the individual attempts to optimize their time and become better, more healthy versions of themselves, chasing the good life (Sutton, 2020: 20).

Importantly, a primary method or selling point for digital detox businesses is promising clients that they can attain a more authentic presence in the world by disconnecting (temporarily). The companies attempt to facilitate experiences of presence with ourselves, with others, and with our temporal and physical surroundings. The three types of presence are all ways of taking control over our attention and carefully minding what we attend to. In this way, the digital detox business reflects the societal concern about the contemporary *crisis of attention* as described by Cook (2018). The crisis of attention is a growing fear that the human capabilities of sustained attention are harmed by technological advancements, as people spend increasing amounts of time on social media or other online rabbit-holes (Cook, 2018). Movements such as mindfulness and meditation see themselves as combatting the crisis of attention, and fear that without efforts to encourage deep, prolonged attention, the human capacity for creativity and intelligence may be dwindling (Wu et al, in Cook, 2018). Cook writes that the 'renaissance of attention' is most likely to come from these practices of self-regulation (Cook, 2018), of which digital detoxing is a prime example.

As we argue, while these companies do not in themselves form a market or an economy as such, they do however point towards some potential future trends where 'the digital' is increasingly being promoted by professionals in the mindfulness and meditation world as a symbolic negation of what is deemed morally and ethically good. As such, selling the promise of presence, as we phrase it, is made possible by such companies through the underlying assumption that the digital is equated with non-presence, in other words, absence from our own minds, each other, and the 'real' world.

Note

1 All names of companies and their owners have been pseudonomized. The pseudonyms used are chosen in accordance with the meaning and message behind the original company names.

Acknowledgement and funding

The research underpinning this chapter has been conducted as part of the ERC funded project *DISTRACT: The Political Economy of Distraction in Digitized Denmark*, project No. 834540. The project has received ethical clearance from the European Research Council, and from the Ethical Committee at the Faculty of Social Sciences, University of Copenhagen.

References

Albris, K., Fast, K., Karlsen, F., Kaun, A., Lomborg, S., and Syvertsen, T. (eds) (2024) The Digital Backlash and the Paradoxes of Disconnection. Nordicom.

Bagger, C. and Lomborg, S. (2021) 'Overcoming Forced Disconnection: Disentangling the Professional and the Personal in Pandemic Times', in A. Chia, A. Jorge, and T. Karppi (eds) *Reckoning with Social Media: Disconnection in the Age of the Techlash*. Rowman & Littlefield.

Baumer, E.P.S., Adams, P., Khovanskaya, V.D., Liao, T.C., Smith, M.E., Sosik, V.S., and Williams, K. (2013) 'Limiting, Leaving, and (Re)Lapsing: An Exploration of Facebook Non-Use Practices and Experiences', *Proceedings of the SIGCHI Conference on Human Factors in Computing Systems – CHI '13*, 3257. ACM Press.

Borup, J. and Jensen, H.J.L. (2015) 'Tema: Mindfulness', *Religionsvidenskabeligt Tidsskrift*, 61(August): 3–4.

Chia, A., Jorge, A., and Karppi, T. (eds) (2021) *Reckoning with Social Media: Disconnection in the Age of the Techlash*. Rowman & Littlefield.

Coker, B.L.S. (2013) 'Workplace Internet Leisure Browsing', *Human Performance*, 26(2): 114–125.

Cook, J. (2016) 'Mindful in Westminster: The Politics of Meditation and the Limits of Neoliberal Critique', *HAU: Journal of Ethnographic Theory*, 6(1): 141–161.

Cook, J. (2018) 'Paying Attention to Attention', *Anthropology of this Century*, 22.

Fish, A. (2017) 'Technology Retreats and the Politics of Social Media', *TripleC: Communication, Capitalism & Critique*, 15(1): 355–369.

Jespersen, M.H., Albris, K., and Willadsen, H. (2025) 'Deconstructing Screen Time: The Connections Between Digital Use, Dissatisfaction, and Disconnection', *Computers in Human Behavior Reports*, 18.

Johnson, N.F. (2020) 'Temporal Digital Control: Theorizing the Use of Digital Technologies to Provide a Temporal Autonomous Space', *Time & Society*, 29(3): 794–812.

Jorge, A. (2019) 'Social Media, Interrupted: Users Recounting Temporary Disconnection on Instagram', *Social Media + Society*, 5(4): 1–19.

Karppi, T., Chia, A., and Jorge, A. (2021) 'In the Mood for Disconnection', *Convergence*, 27(6): 1599–1614.

Lyngs, U., Lukoff, K., Slovak, P., Seymour, W., Webb, H., Jirotka, M., Zhao, J., Van Kleek M., and Shadbolt, N. (2020) '"I Just Want to Hack Myself to Not Get Distracted": Evaluating Design Interventions for Self-Control on Facebook', *Proceedings of the 2020 CHI Conference on Human Factors in Computing Systems*. ACM: 1–15.

Malmlund, J., Suortti, J-P., Behnk, F., and Klein, A.B. (2018) 'A Mobile World: A Study on Nordic Mobile Consumer Behaviour', *Deloitte Global Mobile Consumer Survey 2018*, [online] Available from: https://www.mediebedriftene.no/siteassets/dokumenter/rapporter/deloitte_gmcs-report-2018.pdf [Accessed 3 June 2021].

Merriam-Webster. nd 'How Phone Snubbing Became Phubbing', *Merriam-Webster*, [online] Available from: https://www.merriam-webster.com/words-at-play/phubbing-words-we%27re-watching [Accessed 13 September 2021].

Pedersen, M.A., Albris, K., and Seaver, N. (2021) 'The Political Economy of Attention', *Annual Review of Anthropology*, 50(1): 309–325.

Portwood-Stacer, L. (2013) 'Media Refusal and Conspicuous Non-Consumption: The Performative and Political Dimensions of Facebook Abstention', *New Media & Society*, 15(7): 1041–1057.

Selwyn, N. (2006) 'Digital Division or Digital Decision? A Study of Non-Users and Low-Users of Computers', *Poetics*, 34(4–5): 273–292.

Sutton, T. (2017) 'Disconnect to Reconnect: The Food/Technology Metaphor in Digital Detoxing', *First Monday*, 22(6).

Sutton, T. (2020) 'Digital Harm and Addiction: An Anthropological View', *Anthropology Today*, 36(1): 17–22.

Syvertsen, T. (2020) *Digital Detox: The Politics of Disconnecting*. Emerald.

Syvertsen, T. and Enli, G. (2020) 'Digital Detox: Media Resistance and the Promise of Authenticity', *Convergence*, 26(5–6): 1269–1283.

Syvertsen, T., Karlsen, F., and Bolling, J. (2019) 'Digital Detox På Norsk', *Norsk Medietidsskrift*, 26(2): 1–18.

UFM (2020) 'KU – Det Digitale Danmark – Uddannelses- og Forskningsministeriet', *UFM.dk*, [online] Available from: https://ufm.dk/forskning-og-innovation/forsk2025/indkomne-indspil/videninstitutioner/universiteter/ku-samlede-indspil/ku-det-digitale-danmark [Accessed 18 September 2021].

Williams, J. (2018) *Stand Out of Our Light: Freedom and Resistance in the Attention Economy*. Cambridge University Press.

8

'I Meditate Because My Mind Has Numerous Tabs Open': Dis/connecting (in) Mindfulness Apps

Linda Kopitz

Mediating mindfulness: a contemporary perspective on an ancient practice

In his introduction to an article about the human relationship to technology, self-proclaimed 'personal growth' expert Ross Stevenson asks: 'Why are so many of us feeling more disconnected from the world and others than ever?' (Stevenson, 2018). Expanding on this feeling of disconnection, Trine Syvertsen and Gunn Enli have argued that digital detoxes, the voluntary disconnection from (mostly) online media, do 'not celebrate the opportunities of new media technologies but instead purvey nostalgia for a past where people had more time on their hands, a more defined space and a less stressful lifestyle' (Syvertsen and Enli, 2020: 1270). Paying attention to the implied link between digital connection and a presumed social disconnection, this chapter understands (dis)connection as a prism to negotiate our relationship to technology as much as to each other – and our environment at large. As one potential answer to address the sensation of both disconnection and distraction, the concept of mindfulness has gained an increasing popularity in recent years. Understood as 'an awareness that emerges by way of paying attention intentionally and non-judgmentally, in the present moment, to the unfolding of the moment-by-moment experience' (Plaza et al, 2013: 2) and 'a mental state achieved by focusing one's awareness on the present moment, while calmly acknowledging and accepting one's feelings, thoughts, and bodily sensations, used as a therapeutic technique' (OED, 2020), mindfulness highlights a re-orientation to one's temporal and spatial presence – and, in this regard, juxtaposes the speed and virtuality popularly associated with digitalization.

While the link to psychological therapy is important, mindfulness has become a lifestyle trend as well. Drawing on data published by the US Center for Disease Control and Prevention between 2012 and 2017, Eliza Barclay and Julia Belluz conclude that 'yoga and meditation, two ancient practices, are now officially the most popular alternative health approaches in the United States, each used by around 35 million adults' (2019). At the same time, promoting mindfulness also represents an economic opportunity – especially for the digital economy. As Sarah Perez summarizes in an article published in *TechCrunch*: 'The millennial obsession with wellness and self-care has led to a booming business for meditation apps' (Perez, 2020). According to a study conducted by digital economists at *SensorTower*, 'booming' in this context refers to the 57.4 million new users who downloaded one of the Top 10 mindfulness apps in the year 2019, leading to a combined revenue of $195 million for the companies offering these apps (Williams, 2020). Considering the challenges of the COVID-19 pandemic and its arguably long-term influence on particularly urban lifestyles, these numbers can only be expected to have grown exponentially in the following years. At the same time, meditation and mindfulness apps appear to be situated at the intersection of two (potentially) contradictory lifestyle trends: an increasing digitalization and technologization of everyday activities on the one hand, and an increasing awareness and criticism of our dependence on digital technologies on the other. Across different contexts, this contradiction has led to a reappraisal and return of offline technologies and analogue practices (cf. Sundén, 2015) as not only fostering healthier habits but also more 'meaningful' experiences. This consequently raises the question: how are mindfulness apps themselves negotiating this ambivalence between digital technologization and analogue idealization?

In their critical discussion, Simone Natale and Emiliano Treré highlight how disconnection practices are deeply embedded in 'dynamics of digital capitalism under the innocuous facade of escape in connection to issues of authenticity, mindfulness and nostalgia' (2020: 628). In other words: disconnecting has long shifted from a practice of resistance and refusal to an opportunity for profit. Disconnection in the context of this chapter therefore not only refers to the practical act of (temporarily) disconnecting from digital technologies, but also the affectively charged and increasingly commodified imaginary of disconnection. In line with the historical attention of mindfulness practices to both temporal and spatial presence – to being in touch with one's self as much as one's surroundings – digital meditation and mindfulness apps simultaneously emphasize and disrupt underlying popular assumptions about technology. To draw out this tension, this chapter explores how three leading mindfulness and meditation apps – *Calm, Headspace*, and *Meditopia* – play with themes of (dis)connection in their advertisements through an emphasis on nature experiences, material engagements, and analogue practices. Using audiovisual advertising, published at the time when these apps first received widespread attention and international popularity, as a starting point, this chapter underlines the interconnected, international, and intertwined negotiations of cultural imaginaries of mindfulness and mindful living. While computer studies and psychological research have shown an increasing interest in the design, use, and effects of mindfulness apps, the broader narrative strategies employed by the companies creating and promoting these apps have so far been overlooked. Particularly in light of increasing criticism 'that non-mediated periods and spaces are rapidly disappearing, as epitomized in notions of "on-life" and 24/7 connectivity' (Syvertsen and Enli, 2020: 1275) and ongoing cultural concerns about the commodification of attention by digital technologies (cf. Pedersen, Albris, and Seaver, 2021), paying attention to the positioning of apps quite literally promoting mindful awareness is critically important.

For instance, the marketing language of *Calm*, the leading mindfulness app worldwide, positions the app precisely as a strategy to counter the distraction problem of digital culture: 'Most apps are designed to steal your attention. The *Calm* app was designed to give it back' (Calm, 2020). Bringing the testimonial videos published on Calm's corporate website and YouTube channel into dialogue with videos published by the company's main competitors *Headspace* and *Meditopia*, I propose that mindfulness apps address an inherent criticism of an increasingly technologized, urban lifestyle by framing the app experience as natural, material, and analogue – in other words, by arguing for digital disconnection through technological connection. Linked to historical, scientific, and cultural understandings of mindfulness, the three narrative themes I draw from these advertisements – the natural, material, and analogue – provide a notion of validation and contextualization for digital meditation practices. At the same time, by interweaving the use of mindfulness apps with visual references to established 'healthy habits' (nature experiences, physical connections, and analogue mindfulness practices), the advertisements contribute to making a clear distinction between the use and non-use of technology nearly impossible. Rather than reproducing binary divisions between the natural and the technological, the material and the immaterial, the digital and the analogue, I argue that meditation and mindfulness apps invite us to develop a more nuanced approach to account for the complex unfolding and entangling of these categories within one another – and complicate conceptualizations of an 'analogue idyll'.

'Mindfulness is key to my growth': introducing the 'Calm Stories'

Although the research presented in this case study is limited in scope, the audiovisual narratives discussed here can be understood as representative of a larger framing of meditation and mindfulness apps at the intersection between connection and disconnection. Published in August and September 2019, the YouTube video playlist 'Calm Stories' features four testimonials – John, Alison, Ajani, and Celeste – whose story arc follows the same principle. The videos begin with a shot of the testimonials acknowledging a personal challenge, ranging from emotional break-ups to professional pressures and 'random panic attacks'. As the story shifts from this confessional mode to the search for a solution (read: beginning a mindfulness practice), the camera perspective shifts as well, interweaving interview shots with close-ups of the testimonials and their surroundings. Countering the popular assumption of an urban life increasingly disconnected from our own selves and our environments, these visual shifts play an increasingly important role in entangling technology and mindfulness. Once both the problem and mindfulness as a solution are set up, *Calm* is introduced via a close-up of the app on the testimonial's phone, followed by the selection of one specific exercise and the enactment of the mindfulness practice. Mirroring the introduction, the videos end with an interview perspective in which the testimonials not only highlight their names, but also quite specifically their professional backgrounds. As the length of the videos varies from 2:32 minutes to nearly 4 minutes, so does the length of these different narrative stages. Nonetheless, the overall story arc remains the same across all four testimonial videos. Within the broader marketing landscape of meditation apps, these testimonial videos take on an interesting position: on the one hand, they are representative of so-called 'typical person' testimonials, which involve 'an unknown person, who is presumably representative of the target market, endorsing the product' (Martin, Wentzel, and Tomczak, 2008: 29). On the other hand, the videos are shot as interviews with a certain documentary flair, retelling the experiences of 'real' users. With a language of 'members' and 'personal stories' (Calm, 2019a), these videos emphasize the authenticity – and transferable nature – of the narratives. The selection of testimonials, then, can be seen as representative of both the intended target audience and their potential reservations to using mindfulness apps. By framing the testimonials as users who 'didn't think they were necessarily the kinds of people who could be "good" at meditation' (Calm, 2019b) in an accompanying blog post on the *Calm* website, the app is positioned as a fitting digital solution even for mindfulness sceptics.

Independent of their position between advertisement and information, Brett Martin, Daniel Wentzel, and Torsten Tomczak (2008) claim that the persuasiveness of testimonials depends on three dimensions: expertise, trustworthiness, and source–receiver similarity. By using testimonials who are representative of a target audience, have relatable explanations for their turn to mindfulness practice and emphasize their years of positive experience of using the app, the videos discussed here check all three premises. *Calm* expanded the series in the summer of 2021, featuring testimonials by Cuban-born American singer Camila Cabello (Calm, 2021a) and Canadian musician Shawn Mendes (Calm, 2021b). While the use of celebrities appears to deviate from this checklist at first glance, the emphasis on 'their' struggles just being like 'our' struggles and the placement on the same playlist closes this distance. Through their personal endorsements, John, Ajani, Celeste, and Alison – as well as Camila Cabello and Shawn Mendes – encourage the adoption of mindfulness practices through apps as an everyday habit, which connects to Joseph Bayer and Robert LaRose's discussion of technological habits, 'cues, contexts, and outcome contingencies created by emergent technologies moderate habit formation, performance, and change' (Bayer and LaRose, 2018: 118). As such, both using and explicitly not using digital apps could be understood as 'tech habits' in relation to mindfulness. The following sections will explore how references to nature, materiality, and the analogue throughout the testimonial videos function as narrative tools to blur the distinction between the use and non-use of technology.

While emphasizing the unique position of the *Calm* testimonials, this chapter brings the videos into a productive dialogue with further examples from competing apps to underline the prevalence of these narrative strategies. In line with *Calm*, the mindfulness app *Meditopia*, developed by a German start-up, promises 'less stress and anxiety, better sleep, more peace'. With a particular emphasis on individualization and personalization, the company's YouTube channel functions as an exemplary overview of the different types of practices offered within the app – in particular breathing techniques and meditations based on natural sounds. The emphasis on a visual language of 'wild' landscapes – lush forests, calm seas, eternal green – within both the app and accompanying advertising materials of both *Calm* and *Meditopia* allows for a productive comparison with *Headspace*, the third meditation and mindfulness app drawn on for this chapter. With more than 70 million downloads across all platforms at the time of writing,[1] *Headspace* can not only be considered the main competitor of *Calm* but also takes a unique approach in visualizing mindfulness practice – with an emphasis on colourful animations rather than 'natural' landscape photography. Despite the distinct differences between these three companies (in terms of global reach, positioning, content, visual language, and so on), their promotional videos adopt similar strategies in framing mindfulness and meditation apps by simultaneously conjuring and complicating ideas of an 'analogue idyll'.

Nature: disconnecting by connecting to nature

In an increasingly urbanized and connected world, 'the feeling of being "in the middle of nowhere" is becoming obsolete', as artist and cultural critic Dani Ploeger (2020) has shown with his performative intervention against the ubiquity of GPS mapping and tracking. Yet – or maybe particularly for this reason – the yearning for this place of 'nowhere', of wilderness, continues to saturate our popular imagination. Displacing this longing of an untouched, unfound nature into the urban – and in the process infusing urban lifestyles with 'naturalness' – appears to be one of the key strategies adopted by meditation and mindfulness apps in their positioning. When the 'Calm Story' first introduces John, he is surrounded by nature – solemnly gazing across a natural lake, surrounded by trees. In contrast to the interview scenes, which take place inside his apartment, John does not address or even acknowledge the camera in these outdoor scenes (Calm, 2019d). Rather, he appears to be unaware of anything but the beauty of nature around him, the slow waves on the lake, the rustling leaves in the trees, the clean air. This full immersion into nature – and the positive benefits on both the mind and the body – have been the frequent focus of psychological and medical research. For instance, Terry Hartig et al (2014) draw on previous studies in psychology to conclude that 'for a person experiencing acute stress, nature contact can very rapidly evoke positive affect, which in turn blocks negative thoughts and feelings and fosters reduction of physiological activation' (Hartig et al, 2014: 217). In the digital disconnection study by Trine Syvertsen and Gunn Enli previously mentioned, disconnecting from technology and connecting to nature were frequently intertwined by the respondents: nature is framed as a place 'off the grid', a place 'where the wifi doesn't connect', and, as such, becomes a refuge from technologization. I argue that this popular understanding of technology and nature – connection and disconnection – as polar opposites is precisely why the testimonials for *Calm* so explicitly reference nature: Rather than removing the 'digital' from nature experiences, the 'natural' begins to infuse digital experiences. What Jean Baudrillard (1994) refers to as hyperreality – a state where we cannot distinguish between reality and representations of reality – here functions as a starting point to (re)define a changing relationship between nature and technology, disconnection and connection. In other words: 'being in nature' becomes something to be simulated, experienced at a distance in and through the affordances of technology. Instead of requiring a disconnection from technology, a stepping away from urban and urbanized lifestyles, the grounding, calming, and connective potential of an analogue, natural idyll is framed as accessible via digital meditation.

Following longstanding assertions that 'digital technologies accelerate a radical dislocation of our experience of both space and time' (Barney,

2004: 33), relocating the experience of mindfulness and meditation into a digital and digitized version of nature functions as a reassertion of the grounded experience of time and space, while at the same time subverting the downsides of digital connectivity. For John's 'Calm Story' (Calm, 2019d), the actual mindfulness practice is correspondingly moved outside, his visual situatedness in the landscape becoming a visual – and strategic – shorthand for the immersive and immediate nature experience the app promises to provide. Still seemingly unaware of the camera, John sits down on a tree trunk – thereby also physically connecting to nature – and begins to meditate. Although the app is not explicitly shown in this outside setting, the placement of the phone right next to John's body on the tree trunk insinuates that his mindfulness practice is connected to *Calm*. In John's meditation, the effect of the nature experience and the effect of the app experience thus become connected – and ultimately indistinguishable. Rather than necessitating an investment of time and effort to cultivate a mindful attention to the moment – a practice quite literally requiring years or even a lifetime of practice – mindfulness thereby becomes increasingly entangled with a notion of instantaneity. As part of *Meditopia*'s 'Sounds of Nature' playlist, both the video 'Rain | Relaxing Sounds of Nature' (Meditopia, 2019a) and the video '2 hours of relaxing lake sounds' (Meditopia, 2021a) similarly not only explicitly draw a connection between nature and relaxation in their titles, but also play with the multi-sensorial potential of nature experiences, accessed immediately in and through digital technologies. With shifting shot distances, 'Rain | Relaxing Sounds of Nature' positions the eponymous sound not just anywhere (and particularly not in the urban), but rather amid the vague green of tree leaves. Connecting the 'natural' sound to a 'natural' environment – particularly one that is represented from shifting perspectives, angles, and distances from the viewer – creates the sensation of being enveloped by nature. While the mobilization of nature and nature sounds has a long history in mindfulness practices and philosophies, there is a significant difference to be noted here: the relationship between human and nature, body and environment, relies on the digital mediation of the app. With this portrayal, mindfulness and meditation apps become entangled in the reconfiguration of our understanding of digital technologies – not as obstacles, but crucial access points for nature experiences. However, the seeming realism of 'Rain' is somewhat of an outlier in the playlist: the other three videos, offering the sounds of a lake, a cave, and a jungle respectively, are visually situated in an idealized, pristine, almost too 'perfect', nature. With adjusted colour schemes and a particular geometry, these videos resemble landscapes somewhere between digital screensavers and video games – arguably suggesting that a digitally edited version of nature is the better one. Returning to Jean Baudrillard's concept of hyperreality (1994) here not only underlines how nature and technology become increasingly difficult to define (and experience) as oppositional. Even more provocatively, the 'simulated', digital, and digitized nature might actually provide the more mindful experience with its absence of potential disadvantages and inconveniences of 'being in nature' – mosquitos, rain, tourists.

In this tension between real and hyperreal, it is particularly interesting to note how the textual overlays are blending with the images (or not). While the *Calm* testimonials avoid any obvious reference to (post)editing processes, the *Meditopia* videos not only feature the company's logo but also a description that visually and aurally reasserts both the 'natural' location of the videos (for instance 'Heavy Rain' and 'Lake' in the examples above). The positioning of the font, as well as the delicate typeface and colour of the font itself, makes the text at times appear more visible, at other times less visible – and yet, the evidence of the digital editing of these nature scenes is always there. This presence becomes even more pronounced by the platform-specific digital overlays: YouTube, for instance, displays a pop-up reminder to like the video and subscribe to the channel at random intervals throughout the length of the video. Following Tania Bucher's suggestion that 'phenomena of all sorts – including algorithms – can be "accessed" via experience and the ways in which they make people feel' (2017: 32), these reminder notifications simultaneously disrupt and emphasize the calming experience of the natural imagery. The video '3 hours of jungle sounds for a better sleep' (Meditopia, 2021b) (part of the same 'Sounds of Nature' playlist on the app's official YouTube channel) takes this tension even further by – somewhat surprisingly – visualizing the 'jungle sounds' not with tropical imagery but rather a road cutting through a coniferous forest. More so than with the previous examples, the sounds and visuals entangle nature and technology, disconnection, and connectivity. Filmed as though through the dashboard of a car speeding along this road, complete with approaching headlights in the distance, the chirping of tropical birds feels strangely misplaced between the looming pine trees. At the same time, the humming and thrumming sounds of the rainforest arguably resemble the low-pitched droning sound of tires moving over asphalt just enough to bridge the seeming juxtaposition – thereby drawing another implicit connection between an idealized, untouched nature and modern lifestyles of consumption.

The blending of nature, technology, and urbanity is also palpable in the *Calm* testimonial videos of Ajani and Celeste (Calm, 2019a, 2019c) – even though both exclusively take place within their respective apartments. This reflects the argumentation of Terry Hartig et al (2014) for an inclusion of 'urban nature' in the psychological study of nature experiences, as this terminology 'admits the presence of nature even in those human environments that some consider the antithesis of the natural' (Harting et al, 2014: 208). Although the two testimonials are entirely filmed inside, nature – at least in the sense of a highly stylized, 'urban nature' living situation – is visually referenced. Both Ajani and Celeste are framed by plants in the interview settings, which in turn become the main focus of attention: through soft-focus close-ups and tracking shots, the plants become detached from their urban environment, instead offering the aforementioned understanding of nature as a place of escape and refuge. Interestingly, this immersion into nature through soft-focus shots is especially prominent when the testimonials talk about the positive effects of their mindfulness practice. As James Monaco (2000: 86) has argued, a shift in focus can also function as a tool to direct the viewer's attention. In this example, the focus shifts away from the urban environment and towards the peace of nature. Through this interweaving of the digital experience of the app with visuals of nature, the calming effect of the latter is arguably transferred to the former – undermining preconceived distinctions between technology and nature, connection, and disconnection.

Touch: the materiality of mindfulness

Lara Franco et al's overview of the benefits of nature experiences (2017) emphasizes the multi-sensoriality of nature, highlighting the importance of not just vision but all the senses for the 'full' calming experience of nature. While the previous section has drawn a connection between images and sounds in the construction of a digital 'nature', the focus on smartphone apps points to another dimension of sensory experience: touch. Digital devices are extremely tactile and haptic, designed to be felt and touched. This goes particularly for smartphones which not only seem to be always within our reach but are notably accessed, used, controlled through touchscreens. Yet, research in media studies has highlighted that popular understandings of digital technologies often focus on their immaterial, almost transcendental, properties (cf. Taylor, 2022). In contrast, analogue and offline practices are imagined as more material – and, by extension, as more 'real'. Building on this recurring discourse, Ross Stevenson (2018) urges his readers to create 'healthier relationships to enable a culture of real connection, not one of disconnection through the realms of the digital world'. In this tension, the attention to materiality as something specifically 'offline' in the context of mindfulness becomes particularly interesting. Moving from the positioning of the body within changing environments to the body itself, the examples discussed here place a particular emphasis on the sense of touch – which, in itself, is connected to wellbeing. Lara Franco et al (2017: 871) argue that touch, a frequently overlooked sense in regards to media, can 'affect people's willingness to comply with requests, create bonds between people and groups, strengthen romantic relationships, and reduce stress as measured by blood pressure and heart rate'. As with nature, the emphasis on the benefits of (human) touch seems to be incompatible with digital technology and technological experience. The 'real connections' that Ross Stevenson (2018) calls for appear to necessitate the sensation of 'real touch', which is in turn taken up by the *Calm* testimonials through the foregrounding of the material and sensorial.

On the one hand, each testimonial video features a close-up scene of the testimonial opening the app, scrolling, and ultimately selecting a meditation. Talking about the connection between touch and consumer decisions, Savannah Shi and Kirthi Kalyanam (2018: 45) have argued that 'the positive impact of touch is typically attributed to the increased level of physical tangibility, mental tangibility, and specificity tangibility'. Visualizing the touch of a relatable testimonial and focusing on the screen to allow the viewer to actually see (and later recognize) the app could then be understood as basic marketing techniques. However, I would suggest that these haptic interactions also highlight the physical touch, the physical connection, between the testimonials and the mindfulness app. Contrasting popular imaginaries of the digital as immaterial, the videos discussed here foreground the tactility of the app – in a way 'rematerializing' the digital. Rendering the app interface tangible (rather than immaterial) highlights smartphones and tablets as more than portals to virtual online spaces, but as material presences themselves. In other words, touching digital devices brings us 'in touch' with mindfulness. In this regard, the videos are representative of a sensorial trend in advertising, as Timothy Malefyt (2012: 352) claims: 'marketing to emotions through contact with the senses is thought to bypass issues of reason, diversity, and difference' – figuratively 'touching' consumers across target groups. Extending this into the digital domain allows for an exploration of how the sense of touch, the haptic experience, is highlighted in the marketing of digital products and services often thought to be immaterial. Following the 'broader tendency toward transdigital modes of using and sensing media in contemporary media landscapes' (2015: 136) suggested by Jenny Sundén, of users seemingly effortlessly shifting between analogue and digital in their engagement with different media, mindfulness practices entangle ideas of the analogue and the digital precisely through their materiality. The attention to the body and immediate surroundings in analogue meditation practices – for instance by placing one's hands on the chest or upper abdomen to feel the movement of the breath in calming breathing exercises – becomes shaped through the digital tactility of these mindfulness and meditation apps – for instance by accessing the timer for a breathing exercise through the app's interface. This displacement of different forms of touch further underlines the critiques raised against the conscious attempts of platforms and companies promoting and selling 'digital well-being' to increasingly position themselves 'as both the perpetrator and solution of issues concerning the digital society' (Beattie and Daubs, 2020). Particularly with questions of disconnection, both the problem and its apparent solution appear to be inherently entangled with the digital.

Interestingly, the interaction with the app does not remain the only close-up of touch. In the general format of the 'Calm Stories', all testimonials are portrayed alone – the interviewer is neither heard nor seen, nor do the testimonials directly address the 'viewer'. Highlighting touch – although non-human – might in this regard also be a narrative tool to emphasize connectivity and counter a sense of loneliness and isolation frequently linked to the 'downsides' of digital media. We see Ajani taking photographs of an unseen subject, Celeste and Shawn Mendes making tea (an act that is frequently referenced as something calming in itself) and Alison stretching and touching her own body, with the video lingering on her manicured fingers lightly resting on her feet. Writing on film, Laura Marks (2000) proposes a theory of 'haptic visuality', a visuality that approximates and functions like the sense of touch. Through visions of touch, the distinction between technological connectivity and physical disconnection becomes increasingly blurred – and with it an understanding of which hapticity is desirable and, by extension, 'real'.

If 'content is everywhere and nowhere in particular' (Cave, 2008: 4) in the age of digitization, emphasizing touch can also be understood as a localization of sorts: a physical connection to the present moment in time and space that is important for mindfulness practices. At the same time, during the actual meditation and mindfulness practice, 'digital' touch is specifically excluded: all testimonials are shown selecting a meditation via the app but then placing their phones next to them – the practice itself, once started, does not require holding, tapping, connecting to the device. The explicitness of this movement could be understood as a visual counter to frequent criticism of tech habits as potentially addictive: by replacing 'digital' touch with physical touch when needed, the testimonials are expressive of the neoliberal principle of practising sufficient self-regulation (Bayer and LaRose, 2018: 116) as part of habitual device use. Switching back and forth between these forms of digital and analogue touch in the marketing of – crucially – digital apps, follows Urs Stäheli and Luise Stoltenberg's (2024: 1057) emphasis that 'disconnection has become a marketable entity, promising to help us step out of digital networks, at least for a while'. Here, positioning mindfulness as bringing together multiple senses – vision, sound, and touch – all engaged with and through the materiality of the digital device functions as a counterbalance to notions of the artificial, and specifically the digital, as distanced and distancing.

In her research on voluntary disconnection from technology, Ana Jorge (2019: 2) argues that 'in a mediatized world and hyperconnected age, the distinctions between media/non-media or digital/nondigital are complicated'. By highlighting different forms of touch as – literally – connected, the 'Calm

Stories' further challenge these distinctions: a physical connection between the user and the app is directly followed by a conscious disconnection, which in turn allows the testimonials in the videos to be literally and figuratively 'in touch' with their mindfulness practices – only to ultimately be followed by a return to the digital everyday.

Analogue: highlighting established practices

By beginning his testimonial with 'I meditate because my mind has numerous tabs open' (Calm, 2019a), Ajani not only draws on the language – and criticism – frequently connected to a digitalized lifestyle but also creates a clear juxtaposition between meditation and technology. Building on the previously established links between meditation and mindfulness apps and themes of both nature and touch, the 'closing' of the mental tabs appears to require an even more thorough disconnection from digital technologies. Confronted with the demands of a constantly connected society, Ajani's testimonial resonates with what Jenny Sundén calls a 'desire' for the analogue, 'a longing for an *idea* of the past, for an imaginary time made real through analog practices in the present' (2015: 130). Mindfulness, understood as 'a skill that can be obtained using several training techniques' (Plaza et al, 2013: 2), is inherently linked to habits and practice – and a historical understanding of how, through what techniques and tools, to build this skill. The 'Calm Stories' draw on established, and notably analogue, tools for building mindfulness habits to emphasize this understanding. Alison, for instance, is portrayed in her 'calmest' moments not interacting with the app, but rather practising other, analogue forms of mindfulness. We see her journalling, meditating, and practising yoga, as she is also a self-proclaimed enthusiastic yogi (Calm, 2019b). It should be emphasized here that Alison is not just shown doing 'anything' analogue, but rather following established analogue mindfulness practices linked to both personal development and anxiety therapy (journaling) as well as Buddhism (yoga), in the vaguely referential outdoor setting of a backyard garden. As Immaculada Plaza et al (2013: 2) argue, mindfulness apps 'draw inspiration and take their basic technical features' from Buddhist traditions of meditation. A return to the analogue in the marketing for digital mindfulness apps, also functions as a validation of the efficiency of the (digital) practice through its analogue predecessors.

Similarly, John is shown reading, Ajani practising meditation and yoga, and Celeste drawing Mandalas – again reflecting practices frequently associated with more historical practices and contemporary psychotherapy. Interestingly, the respective smartphones – as access points for the mindfulness apps – remain predominantly out of frame. As a non-tech habit, following these analogue practices would appear to require a digital disconnect at the same time. And yet, it appears to be the use of technology that allows for the building, the establishment, of this analogue habit: through the apps in general but also through connected features like reminders and timers that function as a starting point for a – brief – break away from digital connectivity. Similarly, *Meditopia* suggests that the short breathing techniques that are both part of the app and available via the official YouTube channel 'may be used first thing in the morning to boost your alertness, energy, and focus for the day' (2019b), thereby tying mindfulness practices to questions of performance – of being alert, energetic, and focused in a (most likely) digitalized environment. The switch between inhaling and exhaling in these videos is not only visualized through the words 'inhale' and 'exhale' surrounded by a moving circle but also – crucially – underlined with the sound of a Tibetan gong. With our eyes closed, both the natural landscape, a view across a mountainous landscape partly covered by rolling clouds and the soft light of the rising sun, and the digital interface disappear, leaving us alone with the gong as a historical, recognizable auditory sign for mindfulness and meditation. At the same time, *Meditopia* again draws an implicit connection to other, analogue techniques – in this instance, particularly the 'gong bath', a traditional meditation practice connecting breathwork with the sounds of singing bowls, gongs, and chimes. Through these references, the exact technologies users want to take a 'digital detox' from (as discussed in the self-help literature analysed by Syvertsen and Enli [2020]) become repositioned as tools to actually enhance practices of health and wellbeing.

Headspace adopts a similar-yet-different approach, both in relation to testimonials as advertising tools and established analogue meditation practices as a reference point. While *Calm* highlights the relatability of user stories, *Headspace* launched a video series in 2021 introducing the people behind the voices that guide the different meditations available via the app. This shift to personal stories 'behind the scenes' is a notable contrast to the visual aesthetic of the app interface and its accompanying materials: featuring a recognizable 'cloud' in a variety of situations (and weather conditions), *Headspace* predominantly relies on animated metaphors for mental wellbeing – which in turn underlines the strategic employment of personal stories in the new series. For instance, the video introducing 'Headspace Teacher Kessonga' (2021) not only features Kessonga moving through the aforementioned analogue practices – from touching a singing bowl via lighting a candle to reading a book – but also directly references the book *Full Catastrophe Living* (1990) as a changing point in the teacher's life. First published in 1990, the book by Jon Kabat-Zinn is considered a crucial catalyst for the secular mindfulness movement in the United States, and here forms part of a (seemingly) clear line connecting traditional mindfulness practices via clinical research to *Headspace* as a technological embodiment of both. The emphasis on a personal and professional knowledge charges not only these videos, but also by extension the animated videos written and narrated by the featured protagonists with the affective heritage of established mindfulness practices. *Headspace* makes this connection to established (and inherently non-digital) mindfulness traditions even more central by emphasizing an understanding of meditation 'as both a practice rooted in ancient history and a topic of modern science' (Headspace, 2021).

Generally, there is a trend of returning to the 'analogue' past in different media genres, ranging from a return of print subscriptions to the recent increase in sales of vinyl records (see Palm, 2017). Drawing on similar examples, Claes Thorén et al (2019: 329) argue that the current enthusiasm for analogue and hybrid forms of technology is coloured by 'an almost nostalgic return to the early days of the digital revolution where "connection" required much more conscious engagement and adaptation on the part of the user and the technology being used'. The analogue practices in each *Calm* testimonial video are shown following the visual and narrative introduction of the app itself, suggesting that it was precisely the tech habit of regularly using the app which encouraged the testimonials to establish non-tech habits in their daily lives. In line with Anne Kaun and Emiliano Treré's (2020) understanding of media non-use as 'lifestyle politics', Ana Jorge (2019: 3) argues for a conceptualization of voluntary digital disconnection 'as embedded in the everyday lives of individuals, encompassing their negotiations of online and offline uses in place, time, and social context, and thus as a fluid, ambivalent, highly contingent, and reversible process'. Through their emphasis on analogue practices as 'triggered' by the digital practice, disconnection also becomes a 'performative practice' (Jorge, 2019: 4): a recurring and symbolic loop between digital and analogue, modern and ancient, connected and disconnected. In doing so, the advertisements discussed here add meaning to both phases of use and phases of non-use of technology for establishing a more calm, more mindful life. Positioned as a linear (and logical) extension and expansion of established mindfulness practices, the use of technology, specifically mindfulness and meditation apps, functions as a gateway for the non-use of technology, the immersion into analogue and historical mindfulness practices – a peace of mind that could not be accessed without the technological moderation.

Conclusion: Smarter, calmer, healthier?

In their work on tech habits, Joseph Bayer and Robert LaRose (2018: 117) conclude that 'habit is also a powerful predictor of adoption and continuance for a long list of technologies, usually surpassing the strength of conscious intentions'. Encouraging the habitual adoption of mindfulness practices through apps, then, becomes an important marketing objective for the companies producing and promoting these apps. This negotiation of new habits goes hand in hand with what Urs Stäheli and Luise Stoltenberg (2024) refer to as 'analogization': practices and infrastructures not just creating, but also performing analogue environments. As discussed in this chapter, *Calm* interweaves the use of the mindfulness app with visual references to established 'healthy habits' – nature experiences, physical connections, and analogue practices – thereby making a clear distinction between the use and non-use of technology nearly impossible. At the same time, this strategy arguably positions the use of digital mindfulness apps as the easier (and faster) way to achieve mindfulness, as 'some of us meditate for hours, some seconds' (Calm, 2020) with the same apparent results. If we understand mindfulness as a strategy to 'stabilize attention in the present' (Good et al, 2016: 118), mindfulness can then be easily linked to not just focus, but also efficiency. In the same way that digital technologies have long been promoted as increasing efficiency and productivity, here the digital mediation is presented as a means to increase the efficiency of our mindfulness practice – from sending perfectly timed push notifications, to setting time limits on meditation routines, to guiding us to better ways to breathe. Darren Good et al (2016: 115) draw on the results of studies in psychology and medicine suggesting that 'mindfulness appears to have broadly positive impacts on human functioning' to emphasize the 'productivity' of mindfulness in work situations. Rather than offering a sustainable opting out of stressful lifestyles, these digitized mindfulness and meditation practices appear to ultimately lead us towards the same techno-capitalist logics of efficiently functioning bodies and minds. Just ones that breathe better.

Similarly, Simone Natale and Emiliano Treré (2020: 629) argue that the temporary 'disconnective escape' of digital detox camps 'reveals itself as another way to increase productivity, and is thus, completely functional to the capitalist status quo'. This resonates with the professional careers portrayed in the testimonials: John, a 'dad, professor, DJ and an artist', Ajani, a former 'hardcore workaholic' and 'photographer, director, producer and ultimately, a visual artist', Celeste who 'organizes events', and Alison, 'a mom, a daughter, an auntie, a yogi and a researcher' are representative of a society focused on increasing productivity and (professional) performance as well as juggling multiple roles and identities. The inclusion of Shawn Mendes and Camila Cabello, both recording, touring, and overall highly successful musicians, as additional 'Calm Stories' only further underlines this link between mindfulness and success – especially with both of them weaving their mindfulness practice into their creative labour. 'Mindfulness is key to my growth not just as a musician, but as a person overall' (Calm, 2021b), Shawn Mendes assures us. Similarly, *Headspace* features a playlist quickly introducing meditators and their initial reason for taking up a mindfulness practice, tellingly highlighting the experiences of 'Suveg the Scientist' and 'Clara the Coder' (Headspace, 2017a, 2017b).[2] Mindfulness, afforded and accessed via digital technologies, thereby becomes increasingly tied to the achievement and upholding of 'success'. As part of a neoliberal logic, the idea of unfolding one's potential, both personal and professional, necessitates a constant investment of time and energy. In an ironic twist also mentioned by Simone Natale and Emiliano Treré (2020), the connection to digital technologies is promised to function as a shortcut to experiences of disconnection. While *Meditopia*'s emphasis on time and timers, both via visual and auditory cues, might be the most extreme example of promising this 'shortcut', the notion of an instant mindfulness – fast-tracked, effortless, immediate – is also palpable in the narratives of *Calm* and *Headspace*.

Frequently named on lists of apps that will 'help you get smarter, calmer and healthier in just a few taps on a smartphone' (Stieg, 2020), the framing of digital mindfulness apps appears to mirror – and digitalize – the ideologies associated with the 'Human Potential Movement' in the 1970s: detaching meditation and mindfulness from its religious origins, and instead highlighting the benefits of these practices for the performance and efficiency of the neoliberal subject. While undoubtedly playing into performance culture, the visual and verbal language drawn on by meditation and mindfulness apps constructs an idea of mindfulness beyond recurring critiques of technologization. Through their emphasis on the natural, the material, and the analogue, mindfulness apps like *Calm*, *Meditopia*, and *Headspace* challenge established binaries between connection and disconnection – and complicate the concept of an 'analogue idyll'. First, by referencing 'natural' landscapes and 'natural' soundscapes, the benefits of nature experiences are transferred onto digital and digitized versions of nature as experienced in and through the app. In their mediated perfection and in the absence of other distractions, other people, other realities, this augmented, digital nature arguably becomes more authentic, more 'real', than analogue environments. Second, by highlighting the hapticity of engaging with mindfulness and meditation apps, the advertisements discussed here counteract popular assumptions of the digital as immaterial. Switching between touching technological devices and touching non-technological objects – and bodies – mindfulness becomes rematerialized, even when accessed in and through an app. And finally, by establishing a link between digital and analogue habits – encouraged, tracked, and timed through technology – the use of meditation and mindfulness apps is becoming part of a line of established techniques and tools to achieve a more mindful way of life. Despite their apparent juxtaposition to technologization and digitalization, the natural, the material, and the analogue are folded back into the experience of digitally enabled mindfulness. In other words: achieving this analogue idyll of a natural, material, and analogue mindfulness experience in the everyday is linked – and continues to be linked – to the digital in the testimonials discussed here. Retreating from technology is only possible through technology, mindfulness apps such as *Calm* tell us. We need the digital to be able to return to and enhance the pleasures of the analogue – which in turn provide the mindfulness to live 'calmer, smarter and healthier' in a performance-oriented, technologized society.

Notes

1 Both Calm and Headspace list their current download numbers on their website – 100 and 70 million across platforms respectively (at the time of writing in 2022). While these numbers should be taken with the necessary grain of salt, they are nonetheless an indicator of the digital phenomenon of mindfulness and meditation apps.

2 The videos, published in 2017, all feature the same title 'Headspace | Meditators | Meet …' followed by the respective profession of the testimonials. Their faces are only shown at the end of the roughly 30-second videos following an animated visualization of their story.

References

Barclay, E. and Belluz, J. (2018) 'The Growth of Yoga and Meditation in the US since 2012 is Remarkable', *Vox.com*, [online] 8 November 2018, Available from: https://www.vox.com/2018/11/8/18073422/yoga-meditation-apps-health-anxiety-cdc [Accessed 18 April 2020].

Barney, D. (2004) 'The Vanishing Table, or Community in a World that is No World', *TOPIA: Canadian Journal of Cultural Studies*, 11: 49–66.

Baudrillard, J. (1994) *Simulation and Simulacra*, trans. S. Glaser. University of Michigan Press.

Bayer, J.B. and LaRose, R. (2018) 'Technology Habits: Progress, Problems, and Prospects', in B. Verplanken (ed) *The Psychology of Habit*. Springer International Publishing, pp 111–130.

Beattie, A. and Daubs, M.S. (2020) 'Framing "Digital Well-being" as a Social Good', *First Monday*, 25(12).

Bucher, T. (2017) 'The Algorithmic Imaginary: Exploring the Ordinary Affects of Facebook Algorithms', *Information, Communication & Society*, 20(1): 30–44.

Calm (2019a) 'Calm Stories | Ajani', *YouTube*, [online] 9 September, Available from: https://www.youtube.com/watch?v=U1JB3jbVl20&list=PLgdxvG3UlbieB3yY9UpLgH6JSlwbfr47Q&index=4 [Accessed 11 March 2020].

Calm (2019b) 'Calm Stories | Alison', *YouTube*, [online] 22 August, Available from: https://www.youtube.com/watch?v=p_y4jDwlXlw&list=PLgdxvG3UlbieB3yY9UpLgH6JSlwbfr47Q&index=3&t=27s [Accessed 11 March 2020].

Calm (2019c) 'Calm Stories | Celeste', *YouTube*, [online] 22 August, Available from: https://www.youtube.com/watch?v=5IRxeb3yQgw&list=PLgdxvG3UlbieB3yY9UpLgH6JSlwbfr47Q&index=3 [Accessed 11 March 2020].

Calm (2019d) 'Calm Stories | John', *YouTube*, [online] 22 August, Available from: https://www.youtube.com/watch?v=_lzltFbGtk8&list=PLgdxvG3UlbieB3yY9UpLgH6JSlwbfr47Q&index=1 [Accessed 11 March 2020].

Calm (2020) 'Calm Stories – Calm Blog', *Calm.com*, [online] Available from: https://www.calm.com/blog [Accessed 18 April 2020].

Calm (2021a) 'Camila Cabello x Calm: Breathe Into It', *YouTube*, [online] 17 May, Available from: https://www.youtube.com/watch?v=0503k9zCYAk [Accessed 20 February 2022].

Calm (2021b) 'Shawn Mendes x Calm: Breathe Into It', *YouTube*, [online] 17 May, Available from: https://www.youtube.com/watch?v=YUWaJnQY8dA [Accessed 20 February 2022].

Cave, D. (2008) '"Born Digital" Raised an Orphan?: Acquiring Digital Media through an Analog Paradigm', *The Moving Image*, 8(1): 1–13.

Franco, L.S., Shanahan, D.F., and Fuller, R.A. (2017) 'A Review of the Benefits of Nature Experiences: More Than Meets the Eye', *International Journal of Environmental Research and Public Health*, 14(864): 1–29.

Good, D.J., Lyddy, C.J., Glomb, T.M., Bono, J.E., Brown, K.W., Duffy, M.K., et al (2016) 'Contemplating Mindfulness at Work: An Integrative Review', *Journal of Management*, 42(1): 114–142.

Hartig, T., Mitchell, R., Vries, de, and Frumkin, H. (2014) 'Nature and Health', *Annual Review of Public Health*, 35: 207–228.

Headspace (2017a) 'Headspace | Meditators | Meet Suveg the Scientist', *YouTube*, [online] 30 June, Available from: https://www.youtube.com/watch?v=niQCRkfAdKs [Accessed 20 February 2022].

Headspace (2017b) 'Headspace | Meditators | Meet Clara the Coder', *YouTube*, [online] 28 June, Available from: https://www.youtube.com/watch?v=MWdm-DK7YHI [Accessed 20 February 2022].

Headspace (2021) 'Becoming a Mindfulness and Meditation Teacher: Meet Headspace Teacher Kessonga', *YouTube*, [online] 25 August, Available from: [Accessed 20 February 2022].

Jorge, A. (2019) 'Social Media, Interrupted: Users Recounting Temporary Disconnection on Instagram', *Social Media + Society*, 5(4): 1–19.

Kabat-Zinn, J. (1990) *Full Catastrophe Living: Using the Wisdom of Your Body and Mind to Face Stress, Pain, and Illness*. Random House.

Kaun, A. and Treré, E. (2020) 'Repression, Resistance and Lifestyle: Charting (Dis)connection and Activism in Times of Accelerated Capitalism', *Social Movement Studies*, 19(5–6): 697–715.

Malefyt, T. de W. (2012) 'Touching Consumers, Engaging the Senses in Consumption', *The Senses and Society*, 7(3): 350–355.

Marks, L.U. (2000) *The Skin of the Film: Intercultural Cinema, Embodiment, and the Senses*. Duke University Press.

Martin, B.A.S., Wentzel, D., and Tomczak, T. (2008) 'Effects of Susceptibility to Normative Influence and Type of Testimonial on Attitudes Toward Print Advertising', *Journal of Advertising*, 37(1): 29–43.

Meditopia (2019a) 'Rain | Relaxing Sounds of Nature', *YouTube*, [online] 17 October, Available from: https://www.youtube.com/watch?v=6qOA1KUoh4M [Accessed 20 February 2022].

Meditopia (2019b) 'Energizing breathing techniques', *YouTube*, [online] 16 October, Available from: https://www.youtube.com/watch?v=LgVRctmJdPA [Accessed 20 February 2022].

Meditopia (2021a) '2 hours of relaxing lake sounds', *YouTube*, [online] 7 December, Available from: https://www.youtube.com/watch?v=TWSEmL6s36c [Accessed 20 February 2022].

Meditopia (2021b) '3 hours of jungle sounds for a better sleep', *YouTube*, [online] 5 November, Available from: https://www.youtube.com/watch?v=XX6igMmH1BQ [Accessed 20 February 2022].

Monaco, J. (2000) *How to Read a Film: The World of Movies, Media, and Multimedia: Language, History, Theory*, 3rd edn. Oxford University Press.

Natale, S. and Treré, E. (2020) 'Vinyl Won't Save Us: Reframing Disconnection as Engagement', *Media, Culture & Society*, 42(4): 626–633.

Oxford English Dictionary (OED) (2020) https://www.oed.com/dictionary/mindfulness_n

Palm, M. (2017) 'Analog Backlog: Pressing Records During the Vinyl Revival', *Journal of Popular Music Studies*, 29(4): 1–14.

Pedersen, M.A., Albris, K., and Seaver, N. (2021) 'The Political Economy of Attention', *Annual Review of Anthropology*, 50(2021): 309–325.

Perez, S. (2020) 'Top 10 Meditation Apps Pulled in $195M in 2019, up 52% from 2018', *TechCrunch*, [online] 30 January, Available from: https://techcrunch.com/2020/01/30/top-10-meditation-apps-pulled-in-195m-in-2019-up-52-from-2018/ [Accessed 12 March 2020].

Plaza, I., Demarzo, M.M.P., Herrera-Mercadal, P., and García-Campayo, J. (2013) 'Mindfulness-Based Mobile Applications: Literature Review and Analysis of Current Features', *JMIR mHealth and uHealth*, 1(2): e24.

Ploeger, D. (2020) 'A New Middle of Nowhere', *Ujazdowski Castle Centre for Contemporary Art*, [online] 29 September, Available from: https://u-jazdowski.pl/en/wydarzenia/dani-ploeger [Accessed 20 February 2022].

Shi, S.W. and Kalyanam, K. (2018) 'Touchable Apps: Exploring the Usage of Touch Features and their Impact on Engagement', *Journal of Interactive Marketing*, 44: 43–59.

Stäheli, U. and Stoltenberg, L. (2024) 'Digital Detox Tourism: Practices of Analogization', *New Media & Society*, 26(2): 1056–1073.

Stevenson, R. (2018) 'Disconnection in a Connected World', *Thrive Global*, [online] 22 October, Available from: https://community.thriveglobal.com/disconnection-in-a-connected-world-2/ [Accessed 18 April 2020].

Stieg, C. (2020) '8 Top-rated Apps That Will Make You Smarter, Calmer and Healthier in 2020', *CNBC*, [online] 1 January, Available from: https://www.cnbc.com/2019/12/20/best-apps-that-make-you-smarter-healthier-and-less-stressed-2019.html [Accessed 12 April 2020].

Sundén, J. (2015) 'Technologies of Feeling: Affect between the Analog and the Digital', in K. Hillis, S. Paasonen, and M. Petit (eds) *Networked Affect*. MIT Press, pp 135–149.

Syvertsen, T. and Enli, G. (2020) 'Digital Detox: Media Resistance and the Promise of Authenticity', *Convergence*, 26(5–6): 1269–1283.

Taylor, A.R.E. (2022) 'Cloudwork: Data Centre Labour and the Maintenance of Media Infrastructure', in E. Costa, P.G. Lange, N. Haynes, and J. Sinanan (eds) *The Routledge Companion to Media Anthropology*. Routledge, pp 213–228.

Thorén, C., Edenius, M., Lundström, J.E., and Kitzmann, A. (2019) 'The Hipster's Dilemma: What Is Analogue or Digital in the Post-digital Society?', *Convergence*, 25(2): 324–339.

Williams, K. (2020) 'Mobile Users Spent $195 Million in Meditation Apps Last Year, Up 52% Over 2018', *Sensor Tower Blog*, [online] 31 January, Available from: https://sensortower.com/blog/meditation-apps-2019-revenue-downloads [Accessed 18 April 2020].

Index

References to figures appear in *italic* type; those in **bold** type refer to tables. References to endnotes show both the page number and the note number (33n11).

T

Printed and bound by CPI Group (UK) Ltd, Croydon, CR0 4YY

15/09/2026

14962385-0005